Physical Geography

LABORATORY MANUAL
Fourth Edition

Exercises in Atmospheric & Earth Surface Processes

David Shankman
University of Alabama

KENDALL/HUNT PUBLISHING COMPANY
4050 Westmark Drive   Dubuque, Iowa 52002

Book Team

Chairman and Chief Executive Officer   Mark C. Falb
Vice President, Director of National Book Program   Alfred C. Grisanti
Editorial Development Manager   Georgia Botsford
Developmental Editor   Liz Recker
Prepress Project Coordinator   Sheri Hosek
Prepress Editor   Charmayne McMurray
Permissions Editor   Colleen Zelinsky
Design Manager   Jodi Splinter
Designer   Suzanne Millius
Senior Vice President, College Division   Thomas W. Gantz
Managing Editor, College Field   David Tart
Associate Editor, College Field   Jeanette Laessig

Cover images courtesy of PhotoDisc

Contents

Preface   v
Acknowledgments   vi
About the Author   vii

Lab

 1   **Field Mapping**   1
 2   **Latitude, Longitude, and Time**   9
 3   **Earth-Sun Relations**   21
 4   **Solar Radiation**   35
 5   **Atmospheric Pressure and Winds**   43
 6   **Atmospheric Humidity**   55
 7   **Adiabatic Processes and Precipitation**   65
 8   **Air Masses, Cyclonic Storms, and Fronts**   73
 9   **Weather Analysis**   91
10   **Climate Classification**   99
11   **Topographic Map Interpretation**   121
12   **Trigonometry Applications in Geographical Field Work**   141
13   **Contour Mapping**   151
14   **Soil Properties**   165
15   **Soil Erosion**   183
16   **Vegetation Analysis**   197
17   **Runoff and Infiltration**   221
18   **Stream Discharge**   231
19   **Recurrence Intervals**   247
20   **Aerial Photograph Interpretation and Landform Analysis**   261

Fluvial Processes   265
Arid Landscapes   285
Glaciated Landscapes   301

Appendix A   Trigonometry Tables   319

Appendix B   Diameter and Corresponding Circumference Area   321

Appendix C   Internet Sites   323

Index   327

Preface

This manual is designed for a one or two semester laboratory course in physical geography or environmental science. It will complement most introductory physical geography textbooks. There are 20 laboratory exercises. The first ten exercises focus on earth-sun relationships and atmospheric processes. The last ten exercises deal with mapping, water resources, erosion processes, and landforms. The exercises are self-explanatory and include maps, aerial photographs, and worksheets. The exercises are designed, however, so the lab instructors can, if desired, provide their own maps and aerial photographs to show areas near your location.

This manual is an important alternative to those that are primarily workbooks intended to review classroom lectures. This manual is designed to give students a hands-on, field experience for a more complete understanding of the physical landscape. Many of the exercises in this manual require data collection from the field. However, sample data sets are provided if fieldwork is not practical. In other exercises, data are generated from interpretation of maps and aerial photographs. The data collection, mapping, and analysis required in this manual encourages students to formulate questions and think critically about their observations, to recognize patterns in our physical environment, and better understand the processes that produce them.

The final exercise includes stereo aerial photographs and matching topographic maps. These are divided into sections that show (1) rivers, (2) arid landscapes, and (3) glaciated landscapes. These photographs and maps and the accompanying problems provide reviews for classroom lectures in earth surface processes.

Most lab exercises have an optional Internet component. These include U.S. government sites such as the U.S. Naval Observatory, National Oceanic and Atmospheric Administration, U.S. Geological Survey, and Natural Resources Conservation Service, in addition to university and private sites. Many of the lab exercises utilize regularly updated information or archived data provided by these sites. Other Internet exercises require interpretation of land surface and weather satellite images. A list of all Internet site addresses is in Appendix C.

Acknowledgments

The exercises in this book have been tested in introductory physical geography laboratory sections at the University of Alabama. A special note of thanks is extended to the lab instructors, David Anderson, Judson Englett, Sara Hart, Maria Jose Garcia Quijano, Sara Mace, and Samuel Stutsman whose suggestions and constructive criticisms greatly improved this manual.

Thomas Kallsen, Supervisor of the University of Alabama Map Library, provided time and helpful comments. I am particularly indebted to W. Craig Remington and Daniel P. Royall who provided assistance in many ways. I am grateful for their help.

The figures were prepared by the University of Alabama Cartographic Research Lab.

About the Author

David Shankman completed his doctoral study at the University of Colorado. He is currently Professor of Geography at the University of Alabama. He is also Director of the Environmental Science Program, an interdisciplinary course of study focusing on physical geography, environmental geology, and ecology. He teaches courses in biogeography, geomorphology, and environmental planning.

Dr. Shankman has active research programs dealing with river process and bottomland hardwood ecosystems. He worked for many years in the southeastern U.S. Coastal Plain investigating flood control and modification of alluvial rivers, and the effects of river processes on flood plain forest vegetation. More recently, he has been working in China as the principal investigator of a research project addressing flood control along the Yangtze River. He has published extensively in geography and interdisciplinary science journals.

Name _____ Section _____

Field Mapping

The ability to determine distance, direction, and construct basic maps in the field is an important tool of geographers. Maps, signs, or other navigation tools are available in most areas. But in remote locations, where these tools are not available, you may need to know how to determine your position. The purpose of this exercise is to introduce skills necessary for mapping and for you to construct a map based on your observations and notes taken in the field.

Compass and Measuring Azimuth

Direction will be determined by using a compass. Compass directions are measured in degrees, from 0° to 360°. The four cardinal points of the compass are north (0°), east (90°), south (180°), and west (270°). The four cardinal points of a compass are shown in Figure 1.1. An **azimuth** is a compass direction based on a 360° circle. The needle of a compass points toward the magnetic north, which is not exactly the same as the north pole.

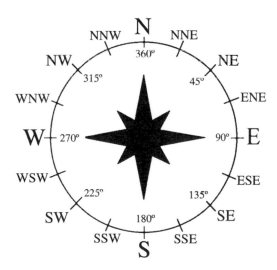

Figure 1.1 Four cardinal points of the compass.

To shoot an azimuth, hold the compass about waist level so that the direction arrow points directly away from you. Rotate the compass housing so that the needle aligns with orienting arrow (or 0° on the compass housing). The degree mark on the housing that is aligned with the direction arrow is your azimuth. Be sure to hold the compass level so that the needle can swing freely.

Two azimuth readings are usually taken to insure the accuracy. After sighting from point A to point B, read an azimuth in the opposite direction, from point B to point A. If the measurements are accurate, there will be a 180 degree difference.

Measure 3 azimuths based on sites given by the lab instructor.

Positions	Azimuth	Back Azimuth	Difference in Degrees
A and B	_____	_____	_____
C and D	_____	_____	_____
E and F	_____	_____	_____
G and H	_____	_____	_____

Pacing

Pacing will be one of the methods you use to measure distance. There are more accurate methods. However, it is important to able to draw rough maps when instruments are not available. Pacing can be moderately accurate except on rough terrain.

The lab instructor has marked a distance of _____ ft or m. Walk from one end to the other counting your paces, or steps. Walk this distance 3 times to determine your average pace.

Number of Paces	Distance per Pace	Average Pace
_____	_____	
_____	_____	
_____	_____	_____ ft or m

Triangulation

Triangulation is a method of locating a point by taking at least two azimuths from known locations, and can be used to determine the distance to a site without direct measurement. To locate an object, measure the azimuth to the object from both ends of a baseline with a known length. The position of the object is determined by the intersection of two azimuths, as shown in Figure 1.2.

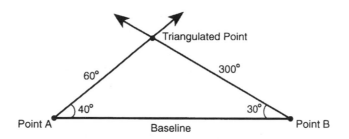

Figure 1.2 Determining the position of a point by measuring azimuths from the endpoints of a baseline.

When triangulating positions, the angles between the baseline and the line of site of the azimuth should be between 30° and 60°. Smaller angles can be used. However, doing so potentially generates large errors, and therefore, great care must be taken.

Position   Azimuth

First Triangulation

Point _____ to point _____   _____°

Point _____ to point _____   _____°

Second Triangulation

Point _____ to point _____   _____°

Point _____ to point _____   _____°

Mapping Exercise

The lab instructor will assign an area to be mapped and will establish a baseline.

1. Measure the azimuth between points designated by the lab instructor and then determine the distance between them by pacing.
2. Triangulate the corners of the area to be mapped (and other features) designated by the lab instructor.
3. Record your notes, measurement, and sketches on the following page.
4. After completing the field survey, plot the features on the graph paper at the end of the exercise using a scale of 1: _____ .

On the graph paper, all vertical lines will be oriented north-south, and the horizontal line will be east-west. Draw the map with a lead pencil, not a pen, in case corrections are needed. Begin with Point 1 and proceed in the order data was collected. The quality of your map depends on accuracy! Use an engineer's scale or ruler for all line work, and a protractor to determine angles.

The map must include a title, scale, date, location, and the azimuth and distance between all points.

Field Notes

From Point	To Point	Azimuth	Paces	Distance	Comments

Latitude, Longitude, and Time

A **geographic grid** is a series of intersecting lines used to locate positions on maps. Most maps are flat and use horizontal and vertical lines. Typically the lines are identified by number and the intersection of lines can be used to indicate a position. The purpose of this exercise is (1) to become familiar with latitude and longitude, the most commonly used geographic grid system to locate positions on the earth's surface, and (2) to understand the relationship between time, longitude, and the earth's rotation.

Latitude and Longitude

Latitude and longitude is a system of imaginary intersecting north-south and east-west lines on the earth's surface.

Latitude is a measure of the distance, or degrees north or south, from the **equator** (Lat. 0°). To determine latitude, there are a series of east-west lines that circle the earth known as **parallels**. The highest latitudes are at the **North Pole** (Lat. 90°N) and **South Pole** (Lat. 90°S) which are fixed points at the earth's axis.

Longitude is a measure of the distance, or degrees, east or west. **Meridians** are north-south lines that are used to measure longitude. The **prime meridian** (Long. 0°) runs through Greenwich, England. All other points of longitude are measures of degrees either east or west of the Prime Meridian, with a maximum of 180° located in the Pacific Ocean. Figure 2.1 illustrates parallels and meridians on the earth's surface.

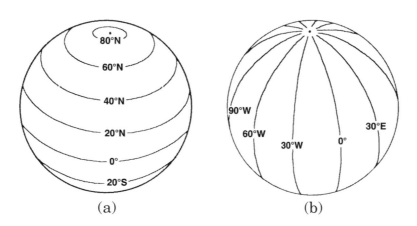

(a)   (b)

Figure 2.1   Parallels (a) and Meridians (b).

On the Figure 2.2 below:

1. Plot and number the following coordinates:
    a. Lat. 44°N, Long. 65°W
    b. Lat. 12°S, Long. 4°E
    c. Lat. 50°N, Long. 60°E

2. Sketch and label the following parallels:
    a. Arctic Circle
    b. Tropic of Cancer
    c. Tropic of Capricorn

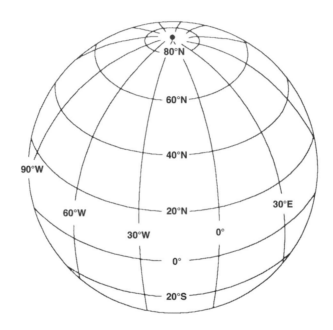

Figure 2.2   Geographic grid coordinates.

The earth is often depicted on flat maps because they are easier to work with than globes. The surface of a sphere, however, cannot be flattened without some distortion of the surface features shown. The Mercator Projection (Figure 2.3) is one of the most common maps of the earth. Note on this map that the meridians are parallel. Actually, meridians are farthest apart at the equator and converge at the poles. Therefore, this map projection causes distortion, particularly at the high latitudes. Still, you can easily use grid coordinates to locate positions on this map.

1. Find your position on the Mercator Projection (Figure 2.3) and determine the geographic grid coordinates.

    Lat. _____ , Long. _____

10

2. The following positions will be given by the lab instructor. Locate and plot each position on the Mercator projection (Figure 2.3).

   Geographic Grid Coordinates   Place Name

   a. _____   _____

   b. _____   _____

   c. _____   _____

   d. _____   _____

   e. _____   _____

3. Shade in the triangular area of the Mercator Projection bounded by:

   Lat. 20°S, Long. 15°E
   Lat. 35°N, Long. 75°E
   Lat. 10°S, Long. 117°E

4. What are the geographic grid coordinates of the positions diametrically opposite the following?

   a. Lat. 18°N, Long. 50°W: _____

   b. Lat. 61°S, Long. 11°W: _____

   c. Lat. 44°, 16°S, Long. 152°,57°E: _____

   d. _____ : _____

11

Figure 2.3   Mercator Projection.

Time

The time of the day is determined by the position of a meridian relative to the sun. It is **noon** on the meridian facing the sun and **midnight** on the diametrically opposite meridian. The earth rotates 360° per day or 15° per hour. The earth rotates west-to-east, and therefore, positions to the east are always later and positions to the west are always earlier. Because the earth rotates, the noon meridian (and the time at all meridians) is constantly changing.

Most of our time keeping uses **standard time zones** that are approximately 15° of longitude wide. All areas within a time zone have the same time of day, which eliminates the need to constantly change times as you move east or west (if you remain in the same time zone). There are 24 time zones, each 15° of longitude and representing one hour of the day. Adjacent time zones differ by one hour. The time zone immediately to the east is one hour later and to the west is one hour earlier.

The time zones are centered on the **central** or **standard meridians** of the zones, which are at 15° intervals beginning with the prime meridian: 0°, 15°, 30°, 45°, . . . 180°. The boundaries of the time zones extend 7½° on either side of the central meridian. The exact boundaries of the time zones, however, are irregular so that they coincide with political boundaries. Figure 2.4 illustrates the world time zones.

Figure 2.4 World time zones. (After U.S. Navy Oceanographic Office, No. 5192.)

14

On the Mercator Projection (Figure 2.5):

1. Mark the central meridians and time zone boundaries for zones in the U.S. Although the boundaries are irregular, assume they are straight in this exercise.

2. Mark the central meridian and time zone boundaries for Greenwich time.

3. Based on your longitude of _____, what is the central meridian of your time zone? _____

4. If the time is _____ in your time zone, what is the time (standard time) at the following meridians?

Longitude	Hour
0°	_____
10°E	_____
_____	_____
_____	_____
_____	_____
_____	_____

5. If it is noon at the following meridians, what is the time in your time zone?

Longitude (noon)	Your Time
0°	_____
33°E	_____
52°W	_____
100°E	_____
_____	_____

Figure 2.5   Mercator Projection.

16

International Date Line

Each new day begins at the **international date line**. When calculating time for different longitudes, remember to change days when crossing the midnight meridian or the **international date line**. The international date line coincides with the 180° meridian which is opposite the Prime (Greenwich) Meridian (Figure 2.6).

Figure 2.6   International date line.

The rules for changing days when crossing the international date line are:

1. When crossing the international date line moving to the west, you move into the next (new) day (Tuesday into Wednesday). All positions west of the international date line until reaching the midnight meridian are part of the new day.
2. When crossing the international date line moving to the east you move into the previous (old) day (Wednesday becomes Tuesday). All positions east of the international date line until reaching the midnight meridian are part of the day.
3. You must change days whenever crossing the midnight meridian or international date line, unless it is midnight on the international date line. In this case, the entire earth is experiencing the same day.

Below are pairs of locations. You are given the time and day for one and are to determine the time and day for the corresponding location. Use the Mercator Projection to find the coordinates of each position.

		Hour	Day		Hour	Day
1.	New York, NY	10:24 p.m.	Wed.	Los Angeles, CA	_____	_____
2.	Denver, CO	2:30 a.m.	Fri.	Toyko, Japan	_____	_____
3.	Greenwich, England	8:30 p.m.	Tues.	Teheran, Iran	_____	_____
4.	Bombay, India	6:18 a.m.	Wed.	Seattle, WA	_____	_____
5.	Toronto, Canada			Cairo, Egypt	_____	_____

Latitude, Longitude, and Time: Internet Resources

Time

http://www.worldtimezone.com/ This site has a world time zone map and shows the current time for each time zone. You can select *Country or Regions* for more detailed maps.

What is the name of your time zone? _____

What is the central meridian (longitude) for your time zone? _____

What is the current time? _____ standard time or daylight time (circle one)

What is the current time in the following locations?

Place Name	Latitude and Longitude	Time
1. _____	_____	_____
2. _____	_____	_____
3. _____	_____	_____
4. _____	_____	_____

http://www.indo.com/distance/ This site can be used to calculate the distance between cities around the world. You can enter the names of the cities or the latitude and longitude of the two places.

http://www.time.gov/ shows official time for time zones in the U.S. and the portion of the earth currently in daylight and darkness. The time given is from the National Institute of Standards and Technology and the U.S. Naval Observatory.

http://timeanddate.com This site is a guide to times and time zones around the world and allows you to view calendars for the year you select.

Map Projections

http://hum.aum.edu.pl/~zbzw/glob.glob1.html includes an extensive number of map projections. It has many global images of interest to earth scientists, such as maps for time zones, global climate and vegetation, and geological features, in addition to satellite images and digital elevation models (DEMs).

Calendars

http://serendipity.magnet.ch/hermetic/index.html This site has a section on *Calendar Studies* with an excellent description and comparison of the Julian and Gregorian calendars.

http://www.ecben.net/calendar.shtml This site gives detailed information for solar and lunar calendars, including Moslem, Jewish, Chinese, Hindu, and other historical and religious calendars.

LAB EXERCISE
3

Earth-Sun Relations

Understanding the earth's movement in space and its changing relationship with the sun is essential for understanding atmospheric and earth-surface processes. The earth's movement directly affects the intensity and duration of solar radiation which accounts for the seasonal changes in temperature and precipitation. Seasonal climatic changes directly affect vegetation, hydrology, soil formation and erosion, ocean currents, among other processes. The purpose of this exercise is to develop an understanding of (1) earth rotation and revolution and (2) changing declination of the sun and how this affects changes in the duration and intensity of solar radiation.

Earth Rotation

Rotation refers to the motion of the earth turning on its axis. The direction of rotation is west-to-east, or counter-clockwise as you look down on the North Pole. A **solar day** is defined as one complete rotation relative to the sun.

While the earth's **angular velocity** is 360°/day (15°/hour), its speed, or **linear velocity**, depends on latitude. The distance around the earth at the equator is 24,900 miles (40,075 km). A point at the equator would move this distance each day. What is the earth's speed at the equator (Lat. 0°)?

_____ miles/hour

_____ km/hour

As shown in Figure 3.1, the distance around each parallel of latitude decreases with distance toward the Poles. Therefore, while every position on earth rotates 360°/day, the distance traveled, and therefore speed, depends on latitude.

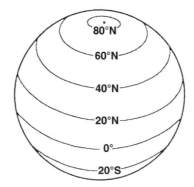

Figure 3.1 Parallels of latitude.

Rotational velocity can be calculated for any latitude by dividing the length of a parallel (distance around the earth at that latitude) by 24, the number of hours in a day. Calculate the rotational velocity for each latitude shown below.

Latitude (N or S)	Length of 1° Longitude		Total Length		Rotational Velocity	
	miles	km	miles	km	mph	kmph
20	65	105	_____	_____	_____	_____
40	53	85	_____	_____	_____	_____
60	35	56	_____	_____	_____	_____
80	12	19	_____	_____	_____	_____

What is the rotational velocity of:

North Pole (Lat. 90° N)? _____

South Pole (Lat. 90° S)? _____

Declination of the Sun

The earth revolves in its orbital path around the sun on the **plane of the ecliptic**. The earth's axis is inclined 23½° to a line perpendicular to the ecliptic (or 66½° from the ecliptic). The earth's axis is in a fixed position relative to the stars, so that the North Pole (Lat. 90°N) always faces the North Star.

As the earth moves in its orbital path, the axis changes its position relative to the sun. (Figure 3.2.) Because of the earth's 23½° axis tilt, the direct ray of the sun, or the latitude on earth where the sun is directly overhead, moves between the **tropic of Cancer** (Lat. 23½°N) and the **tropic of Capricorn** (Lat. 23½°S). The direct ray of the sun, or **declination**, for every day of the year is shown on the **analemma** (Figure 3.6) at the end of this exercise. Note that the direct ray of the sun is farthest north in late June, which is the **summer solstice**, and farthest south in late December, the **winter solstice**. The sun crosses the equator (Lat. 0°) twice each year on the **March** and **September equinoxes**.

Refer to the analemma (Figure 3.6) for the following questions:

1. What is the declination of the sun?

    today: _____ °

    one week from today: _____ °

    one month from today: _____ °

    six months from today: _____ °

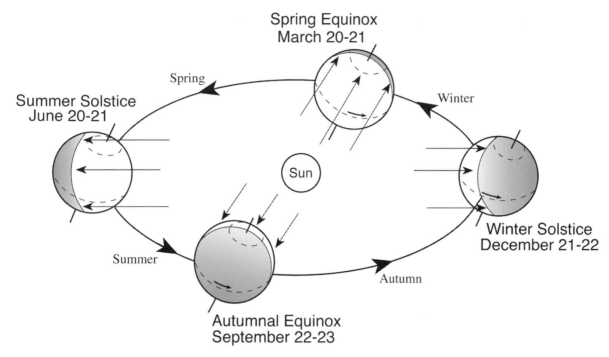

Spring Equinox
March 20-21

Spring

Summer Solstice
June 20-21

Winter

Sun

Summer

Winter Solstice
December 21-22

Autumn

Autumnal Equinox
September 22-23

Figure 3.2   Earth's revolution and axis position relative to the sun.

2.   Between what two dates is the declination of the sun

     moving north? _____

     moving south? _____

3.   Between what two dates is the declination of the sun

     in the northern hemisphere? _____

     in the southern hemisphere? _____

Length of Day and Night

Because the earth is a sphere, one half (the part facing the sun) is illuminated at any point in time. However, because the position of the axis is changing relative to the sun, the amount of daylight, as opposed to darkness, changes at all positions or latitudes, except at the equator which always has 12 hours of daylight and darkness. Figure 3.3 shows the earth-sun relationships on the June (summer) and December (winter) solstices.

Rules for the length of day and night

1.   If the declination of the sun is in the northern hemisphere, the length of day is greater than night for all positions in the northern hemisphere, and length of night is greater than day for all positions in the southern hemisphere.
     Conversely, if the declination of the sun is in the southern hemisphere, the length of day is greater than night for all positions in the southern hemisphere, and length of night is greater than day for all positions in the northern hemisphere.

23

2. If the declination of the sun is in the northern hemisphere, there is progressively longer period of daylight with distance north, until reaching the area receiving 24 hours of daylight.

   If the declination of the sun is in the southern hemisphere, there is progressively longer period of daylight with distance south, until reaching the area receiving 24 hours of daylight.

3. All parallels of latitudes will have either greater periods of daylight or darkness for each 24 hour period except on an equinox when all parallels have 12 hours day and night. The exception is the equator (Lat. 0°) which always has 12 hours of daylight and darkness.

4. As the declination of the sun moves north (from the December solstice to the June solstice) the period of daylight increases in the northern hemisphere and decreases in the southern hemisphere.

   As the declination of the sun moves south (from the June solstice to the December solstice) the period of daylight increases in the southern hemisphere and decreases in the northern hemisphere.

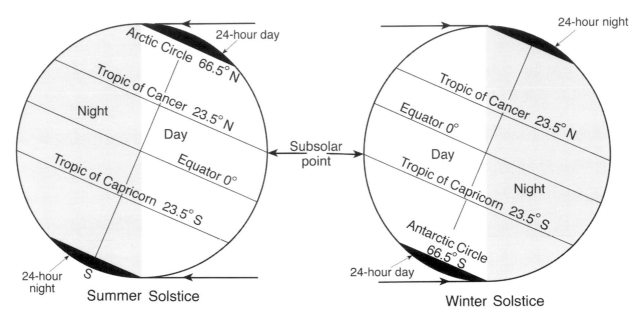

Figure 3.3   Illumination of the earth on the solstices.

Refer to Figure 3.3 to answer questions 1 and 2.

1. Compare the length of daylight at the tropic of Cancer (Lat. 23½N) on the June and December solstices.

2. Compare the length of daylight at the tropic of Capricorn (Lat. 23½S) on the June and December solstices.

24

Questions 3–7 refer to your latitude: _____ ° N or S

3. Today (date: _____) is the period of daylight or darkness longer?

    _____

4. Tomorrow, will the length of day be longer or shorter than today?

    _____

5. On what day of the year will the period of daylight be longest?

    _____

6. On what day of the year will the period of darkness be longest?

    _____

7. During what time of year do you have longer periods of daylight than darkness?

    _____

8. Compare the lengths of daylight today between:

    10°N and 20°N: _____

    20°S and 0°: _____

    _____ and _____: _____

9. Which of the following latitudes:

    Tropic of Cancer (Lat. 23½°N)
    Equator (Lat. 0°)
    Tropic of Capricorn (Lat. 23½°S)

    will have a greater period of daylight on:

    January 1? _____

    May 14? _____

    August 25? _____

    October 1? _____

    _____ ? _____

    _____ ? _____

25

10. On the following date: _____ , is the length of daylight increasing or decreasing at:

Lat. _____ : _____

Lat. _____ : _____

Lat. _____ : _____

Lat. _____ : _____

There are always areas on earth that have either 24 hours of daylight or darkness (except during an Equinox when all parallels have 12 hour days and nights). These are the regions in the high latitudes that do not rotate either into or out of the **circle of illumination** (or surface area exposed to the sun) as the earth turns on its axis. The areas on earth receiving 24 hours of daylight or darkness can easily be determined for any day of the year by referring to the analemma.

First, find the declination of the sun for the desired date and then subtract it from 90 (90 – declination). The resulting value represents the latitudes at and above which there is either 24 hours daylight or darkness. For example, if the declination of the sun is 10°N (90 – 10 = 80), latitudes 80°N and S and all higher latitudes will have either 24 hours daylight or darkness. Since in this case the sun is in the northern hemisphere, all areas ≥ 80°N will have 24 hours daylight on that day. All areas ≥ 80°S will have no daylight.

Since the maximum declination of the sun is 23½°N or S, the lowest latitude to have 24 hours of daylight and darkness at some point during the year is Lat. 66½°N (**Arctic Circle**) and Lat. 66½°S (**Antarctic Circle**) (90 – 23½ = 66½).

1. What areas will receive 24 hours of daylight?

today: _____

one month from today: _____

June Solstice: _____

Date: _____ : _____

Date: _____ : _____

2. What areas will receive 24 hours of darkness?

today: _____

one month from today: _____

December Solstice: _____

Date: _____ : _____

Date: _____ : _____

3. Between what two dates will the following positions have 24 hours of daylight?

North Pole: _____

80°N: _____

70°N: _____

Why is the period of continuous daylight decreasing as you move to a lower latitude?

Altitude of Noon Sun

The **altitude of the sun** is the angle of the sun above the horizon. The sun reaches its highest point in the sky each day at noon. Since the sun is highest in the sky at noon, the intensity of solar radiation is greatest at this time each day. Figure 3.4 shows the altitude of the sun at the solstices and equinoxes.

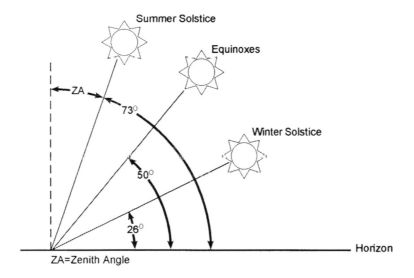

Figure 3.4   Altitude of the noon sun at the solstices and equinoxes at 40°N.

The **altitude of the noon sun (ANS)** will vary depending on latitude. Because the earth is a sphere, the sun can be directly overhead at only one latitude (shown by the analemma) on any day. As the earth moves in its orbital path and the declination of the sun changes, the ANS also changes. If you know your latitude, the ANS can be determined for any day of the year.

27

The ANS at any latitude can be calculated using the following equation:

ANS = 90 – Zenith, or
ANS = 90 – (arc of earth's surface between LP and LS)

where:
LP is your latitudinal position and
LS is the declination of the sun. Therefore,

$$\text{ANS} = \begin{cases} 90 - (\text{LP}+\text{LS}) & \text{if LP and LS are in different hemispheres} \\ 90 - |\text{LP}-\text{LS}| & \text{if LP and LS are in the same hemisphere} \end{cases}$$

Note: If the sun is directly overhead, the LP and LS is the same, and therefore, the zenith angle (or difference between LP and LS) is 0. In other words, if the sun is directly overhead, it is 90° above the horizon:

ANS = 90 – 0
ANS = 90

But, for every degree difference between the LP and the LS, the sun is that many degrees from the direct overhead position. For example, as shown in Figure 3.5, the declination of the sun (LS) on May 21 is 20°N. If you were at 20°N (LP), the noon sun would be directly overhead. Now assume that you are at 13°N (LP), or 7° from the declination of the sun (LS). So, instead of the sun being directly overhead, it is 7° from the straight overhead position. The zenith angle is 7°.

ANS = 90 – (20 – 13)
ANS = 90 – 7 = 83°

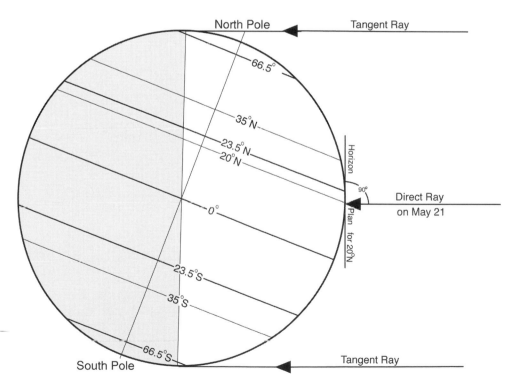

Figure 3.5   Earth-sun relationship on May 21 and July 25.

28

Use your latitude position (LP) for questions 1–5.

Latitude: _____ Today's date: _____

1. The declination of the sun today is _____

2. What is the ANS today at your latitude? _____

3. On what day of the year is the ANS lowest? _____

 What is the ANS on this day? _____

4. On what day of the year is the ANS highest? _____

 What is the ANS on this day? _____

5. What is the ANS on an equinox? _____

6. Today, on what part of the earth's surface does the period of daylight exceed the period of darkness?

7. At what latitudes (in both northern and southern hemispheres) will the sun never be directly overhead?

8. What is the highest and lowest ANS for the following latitudes?

Latitude	ANS	
	Highest	Lowest
_____	_____	_____
_____	_____	_____
_____	_____	_____
_____	_____	_____

Fill in all blanks in the following table:

ANS	LP	LS	Date
_____	10°N	_____	March 28
_____	38°N	_____	October 1
_____	22°S	14°N	_____
_____	0°	2°S	_____
_____	80°N	8°S	_____

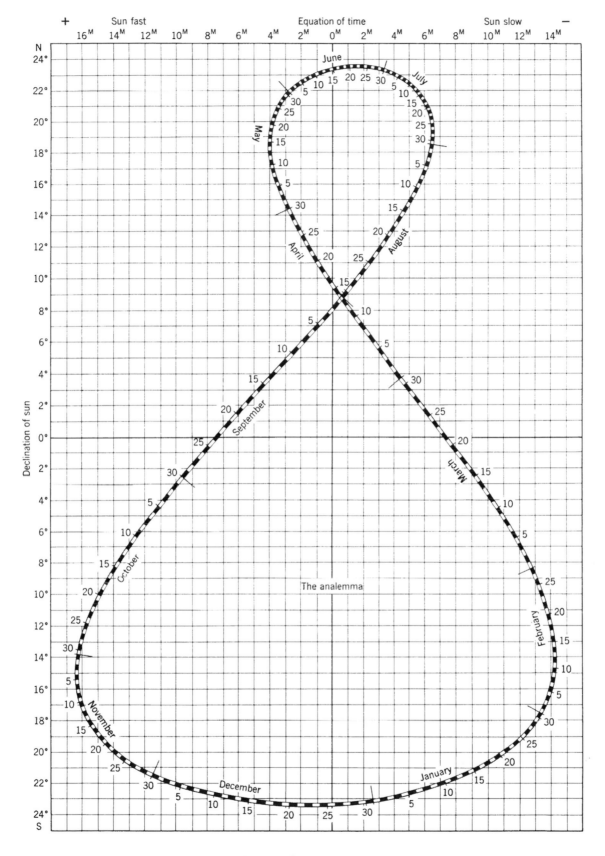

Figure 3.6   **The analemma gives the declination of the sun and the equation of time for each day in the year. (From U.S. Coast and Geodetic Survey.)**

Earth-Sun Relations: Internet Resources

Time for Sunrise/Sunset and Solstices and Equinoxes

http://www.usno.navy.mil/ is the site for the U.S. Naval Observatory. Its purpose is to determine timing and astronomical dates for navigation. Select *Astronomical Applications* and then *Data Services*. The section on *Complete Sun and Moon Data* provides sun and moon rise and set times for any location you select.

Date: _____ Location: Latitude _____ ; Longitude _____

Today, what time is sunrise? _____ sunset? _____

Today, how long is there daylight? hours: _____ minutes: _____

What time is sunrise and sunset, and how long is there daylight on the following dates?

Tomorrow (date: _____)
sunrise: _____ sunset: _____ length of daylight: _____

Day after tomorrow (date: _____)
sunrise: _____ sunset: _____ length of daylight: _____

Two days after tomorrow (date: _____)
sunrise: _____ sunset: _____ length of daylight: _____

One month later (date: _____)
sunrise: _____ sunset: _____ length of daylight: _____

Six months later (date: _____)
sunrise: _____ sunset: _____ length of daylight: _____

Compare sunrise and sunset to a position at the same degrees latitude in the southern hemisphere.

Longitude: _____
sunrise: _____ sunset: _____ length of daylight: _____

You can also select *Earth's Seasons* for the exact time of solstices and equinoxes. Note that the times for these dates use Greenwich Mean Time.

http://www.sunrisesunset.com This is another site that calculates the time of sunrise and sunset for locations and dates you select.

http://nws.noaa.gov/er/box/equinox.html gives the exact time of solstices and equinoxes 1992–2005. Eastern Daylight Time and Eastern Standard Time are used.

Daylight and Darkness on Earth

http://www.fourmilab.ch/earthview/vplanet.html has a world map showing day and night regions of the Earth at the current time.

http://www.worldtimezone.com/ This is another site that shows day and night regions of the Earth at the current time. Select **Sunclock**. Note the relative portions of the northern and southern hemispheres that are in daylight and darkness.

Today, what part of the Earth (what latitudes) has:

A longer period of daylight compared to darkness? _____

A longer period of darkness compared to daylight? _____

What is the relationship between solar declination and periods of daylight vs. darkness?

Name _____ Section _____

LAB EXERCISE
4

Solar Radiation

Equipment

radiometer
thermometer
soil thermometer

Solar radiation is the major heat source for the earth's surface and is the power controlling the earth's weather. Solar radiation also, directly or indirectly, affects almost all earth surface processes, including vegetation, soil moisture, stream flow, weathering processes, and ecosystem functioning. In this exercise, data will be collected from different sites to determine how small scale variations in radiation affect temperature. Also, you will calculate the intensity of solar radiation for different latitudes and topographic positions based on the altitude of the sun.

Because of the differences in the altitude of the sun with different latitudes, the intensity of solar radiation at the surface is highly variable. Also, some of the solar radiation reaching the earth's surface is reflected back to space and does not contribute to surface heating. The reflectivity of a surface, or **albedo** (A), is usually expressed as a percentage and can be defined as:

$$A = \frac{\text{reflected solar radiation}}{\text{incoming solar radiation}} \times 100$$

The albedo of a surface is highly variable. With identical solar angles, dark or rough surfaces, such as those covered by asphalt or vegetation, typically have low albedos. Whereas, light or smooth surfaces, such as those covered by snow or ice, usually have high albedos. Since reflected solar radiation does not heat surfaces, surface albedo can directly affect both surface and air temperature.

Data Collection

Three sites will be designated from which you will collect radiation and temperature data. A radiometer will be used to measure both incoming and reflected solar radiation for each study site. The ratio of these two readings will then be used to calculate albedo as shown in the above equation.

On each site, measure the temperatures at the surface, 1 meter above the surface, and at least 8 cm (about 3 inches) below the surface, except on the black asphalt. The subsurface temperature readings should be taken at the same depth on the remaining sites. Record the temperatures and radiation values on the following table. Then calculate albedo.

	Black Asphalt	Vegetated Surface Full Sun	Shade	_____
Temperature (°F):				
1 m above surface	_____	_____	_____	_____
Surface	_____	_____	_____	_____
Soil		_____	_____	_____
Solar Radiation				
Incoming	_____	_____	_____	_____
Reflected	_____	_____	_____	_____
Albedo	_____ %	_____ %	_____ %	_____ %

Note: It is best to take temperature readings when there is little wind. Even moderate wind causes turbulence and mixing of the air near the surface. Mixing will reduce or eliminate both vertical and horizontal temperature differences that normally occur when conditions are calm and will limit the ability to make useful interpretations from this data.

On Figure 4.1 plot the three temperature readings for each study site. Then, connect the points to create temperature profiles.

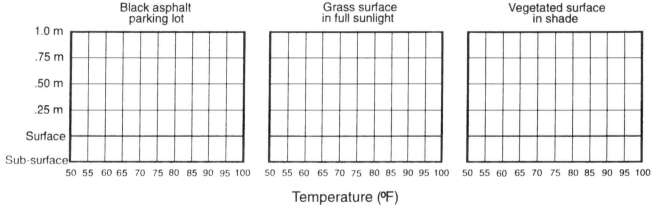

Figure 4.1   Create temperature profiles.

1.   On which site is albedo greatest? Why?

2.   On which site is the temperature difference between the surface and 1 meter elevation greatest? Why?

3. Which surface has the highest temperature and what is the relationship between surface temperature and albedo?

4. Are the soil temperatures similar on the shaded and unshaded sites? Explain.

5. What differences would occur in temperature at each elevation under cloudy conditions?

Intensity of Solar Radiation

The intensity of solar radiation depends on the angle of the solar beam which is determined by latitude and the changing declination of the sun. If the sun is directly overhead, the intensity of radiation is greatest because of its concentration in a small area. As shown in Figure 4.2, with a lower angle, the same amount of radiation is distributed over a larger area and therefore is less intense.

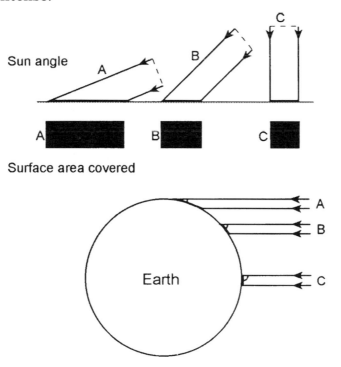

Figure 4.2 Angle of the sun's rays and concentration of radiation.

37

The intensity of radiation for different sun angles is shown in Table 4.1. Radiation intensity for given angles (altitude) of the sun is expressed as a percentage of that of a beam directly overhead (perpendicular to the surface).

Table 4.1   Intensity of Solar Radiation

Angle of Beam	0°	1°	2°	3°	4°	5°	6°	7°	8°	9°
0°	00.0	01.8	03.5	05.2	07.0	08.7	10.5	12.2	13.9	15.6
10°	17.4	19.1	20.8	22.5	24.2	25.9	27.6	29.2	30.9	32.6
20°	34.2	35.8	37.5	39.1	40.7	42.3	43.8	45.4	47.0	48.5
30°	50.0	51.5	53.0	54.5	55.9	57.4	58.8	60.2	61.6	62.9
40°	64.3	65.6	66.9	68.2	69.5	70.7	71.9	73.1	74.3	75.5
50°	76.6	77.7	78.8	79.9	80.9	81.9	82.9	83.9	84.8	85.7
60°	86.6	87.5	88.3	89.1	89.9	90.6	91.4	92.1	92.7	93.4
70°	94.0	94.6	95.1	95.6	96.1	96.6	97.0	97.4	97.8	98.1
80°	98.5	98.8	99.0	99.3	99.5	99.6	99.8	99.9	99.9	99.9

To determine the relative intensity of a solar beam with an angle of incidence of 53 degrees, read down the left-hand column to 50 and then across the row to 3. The relative intensity for an angle of incidence of 53 degrees is 79.9 percent.

For the following problems you must determine the altitude of the noon sun (ANS) for different dates at given latitudes. Refer to the analemma (Figure 3.6) to make this calculation. The latitude of your position is

_____ °N

Use Table 4.1 to answer the following questions.

1. Today, the altitude of the noon sun (ANS) at your latitude is _____ .

   The intensity of solar radiation at noon at your latitude is _____ percent.

2. On what date will maximum intensity of radiation occur at your latitude?

   _____

   What is the altitude of the noon sun (ANS) on that day? _____

   What is the maximum intensity of radiation that occurs on that day?

   _____ percent

3. On what date will minimum intensity of radiation occur at your latitude?

    _____

    What is the altitude of the noon sun (ANS) on that day? _____

    What is the minimum intensity of radiation that occurs on that day?

    _____ percent

On the following table include the ANS and intensity of radiation for the given latitudes.

Latitude	ANS		Intensity of Radiation (%)	
	Highest	Lowest	Maximum	Minimum
0°	_____	_____	_____	_____
10°N	_____	_____	_____	_____
_____	_____	_____	_____	_____
_____	_____	_____	_____	_____
_____	_____	_____	_____	_____
_____	_____	_____	_____	_____

The intensity of radiation over a small area is also affected by topography. In the U.S. the noon sun is always in the southern sky (even though its exact position changes). Therefore, the intensity of radiation is higher on south-facing slopes than on north-facing slopes at the same latitude.

   To determine the intensity of radiation on an inclined surface you must determine (1) the solar angle on flat ground at the same latitude, and (2) the angle of the slope. The solar angle on the ground (SA_g) is calculated by:

$$SA_g = SA \pm \alpha$$

   where: SA = solar angle on a flat surface
           α = slope angle in degrees

Figure 4.3 illustrates how much the intensity of radiation will vary on different facing slopes, even those that are gently inclined.

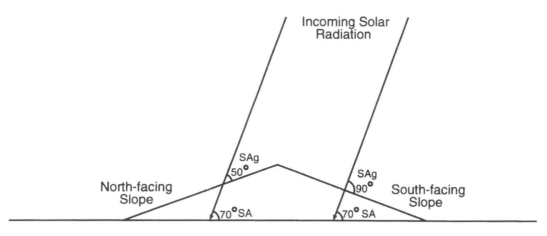

Figure 4.3   Angle of the sun on different slopes.

In the following table, slope angles for north and south facing slopes are given. Complete the table by calculating the intensity of solar radiation at noon on slopes with different angles in your area. Refer to Table 4.1.

Latitude: _____   Today's date: _____   ANS: _____

Slope Angle (°)				Intensity of Radiation (percent)	
north-facing	SAg	south-facing	SAg	north-facing	south-facing
5	____	5	____	____	____
10	____	10	____	____	____
____	____	____	____	____	____
____	____	____	____	____	____

For the following table, dates and latitudinal position (and slope angle where not shown) will be given by the lab instructor. Based on this information, complete the table by calculating the ANS and the intensity of radiation.

Date	Latitude	ANS	Slope Angle (°)		Intensity of Radiation	
			north-facing	south-facing	north-facing	south-facing
_____	_____	____	5	5	____	____
_____	_____	____	10	10	____	____
_____	_____	____	____	____	____	____
_____	_____	____	____	____	____	____
_____	_____	____	____	____	____	____

40

Solar Radiation: Internet Resources

Temperature Anomalies

http://wwwghcc.msfc.nasa.gov/temperature/ NASA maintains the Global Hydrology and Climate Center. The image on this site shows temperature anomalies for the most recent month of record and can be reset for each month during the past two decades.

Below, note how much monthly temperatures were above or below average at your location for the previous six months, beginning with the most recent month on record.

Lat. _____ ; Long. _____

Month   Degrees above (+) or below (-) the monthly average

1. _____   _____ °C

2. _____   _____ °C

3. _____   _____ °C

4. _____   _____ °C

5. _____   _____ °C

6. _____   _____ °C

Note how much monthly temperatures were above or below average at the Antarctic Penninsula (northernmost part of Antarctica), or other position indicated by your lab instructor, for the previous six months beginning with the most recent month on record.

Lat. _____ ; Long. _____

1. _____   _____ °C

2. _____   _____ °C

3. _____   _____ °C

4. _____   _____ °C

5. _____   _____ °C

6. _____   _____ °C

Global Warming

http://www.epa.gov/globalwarming/index.html This is the U.S. Environmental Protection Agency's Global Warming Site. It has information on what is known about recent climate change and discusses predictions for future climate change. This site also discusses emissions of greenhouse gases and possible environmental impacts.

Atmospheric Pressure and Winds

The earth's atmosphere consists primarily of gases held to the earth by gravity. The weight of the atmosphere produces a force on the earth's surface called **pressure**. The standard unit to measure atmospheric pressure is the **millibar** (mb). Average atmospheric pressure is 1013.2mb (equivalent to 14.7lb/sq. inch). This is referred to as **standard sea-level pressure**, however, pressure at the earth surface varies for a given time and place. Pressure is among the most important atmospheric conditions influencing weather. It determines wind direction and speed and strongly affects the probability of precipitation. The purpose of this exercise is for you to become familiar with the relationship of pressure to winds.

Altitude and Atmospheric Pressure

Density of the atmosphere is greatest at the earth's surface (at or near sea-level) and thins with increasing height. If you travel from the earth's surface up to 5.6 km, about one-half of the total mass of the atmosphere will be below you. If you move upward another 5.6 km (11.2 km above the earth's surface), about one-fourth of the atmosphere is above with three-quarters below. Generally, regardless of the beginning elevation, the atmospheric mass decreases at a rate of 50 percent every 5.6 km increase in elevation.

1. Based on the 5.6 km rule, determine the atmospheric mass above and below for each elevation (height above sea-level) in Table 5.1. (Atmospheric pressure will be determined later.)

Table 5.1   Determine Atmospheric Mass and Pressure

Height Above Sea-Level (kilometers)	Percent Atmospheric Mass Above	Below	Atmospheric Pressure (mb)
39.2	_____	_____	_____
33.6	_____	_____	_____
28.0	_____	_____	_____
22.4	_____	_____	_____
16.8	_____	_____	_____
11.2	_____	_____	_____
5.6	_____	_____	_____
0.0 (sea-level)	100	_____	_____

2. Plot the designated values for *percent atmospheric mass* from the above table on Figure 5.1. After plotting these values, *carefully draw a smooth curving line* that best fits these data and shows the changing relationship between elevation and pressure.

Based on the graph you have created on Figure 5.1 answer the following questions.

1. The highest elevation in your state is _____ m.

    What percentage of the atmosphere lies above this elevation? _____ percent

2. The highest mountain in North America excluding Alaska is Mt. Whitney. Its peak is 4418m above sea-level.

    What percentage of the atmosphere is above Mt. Whitney? _____ percent

3. Mt. Everest is the highest mountain in the world. Its peak is 8848m. What percentage of the atmosphere is above Mt. Everest? _____ percent

4. What is the height above which only 5 percent of the atmosphere's mass lies? _____ km

Barometric pressure is the weight of the atmosphere and can be measured at the Earth's surface or at some point above the surface. With increasing elevation there is less atmospheric mass and a corresponding decrease in atmospheric pressure. Only 50 percent of the atmosphere lies above 5.6 km. Therefore, atmospheric pressure at that elevation is 50 percent of 1013.2 mb (standard sea-level pressure), or 506.6 mb.

1. On the right side of Table 5.1 is a column labeled "Atmospheric Pressure." Determine the atmospheric pressure (mb) for each elevation shown.

2. Locate and label atmospheric pressure values on Figure 5.1.

3. Use Figure 5.1 to determine barometric pressure for the following elevations:

    height above which 50 percent of the atmosphere lies: _____ mb

    height above which 5 percent of the atmosphere lies: _____ mb

    height of Mt. Everest: _____ mb

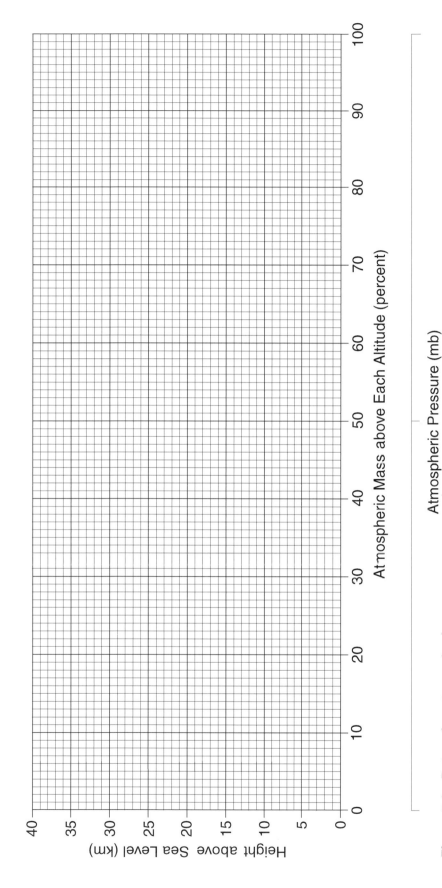

Figure 5.1 Determine atmospheric pressure.

45

Pressure and Winds

Small and large scale differences in pressure from one place to another are caused by differences in temperature at the earth's surface. The differences in pressure result in a **pressure gradient**. Pressure gradients at the earth's surface are the driving force of winds. Where there is a strong pressure gradient (a large difference in pressure over a short distance) winds will be strong. Conversely, a weak pressure gradient will result in weaker winds. **Isobars** are lines on a barometric pressure map that connect points of equal pressure. The strength of a pressure gradient can be shown on a weather map by the spacing of isobars. The closer the spacing of isobars, the steeper the pressure gradient. *The pressure gradient force is always directed perpendicular to isobars, away from high pressure and toward low pressure.*

Figure 5.2 is a small section of a surface map showing atmospheric pressure. A larger pressure map would show circular isobars indicating areas of high and low pressure. On this figure:

1. Indicate areas of highest (H) and lowest (L) pressure.

2. Use an arrow to show direction of the pressure gradient.

3. Indicate where the pressure gradient is strong (s) and weak (w), and where the winds are strongest (st).

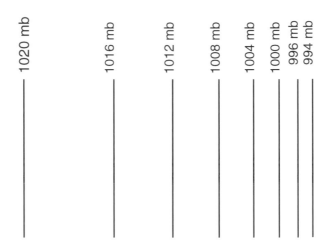

Figure 5.2   Section of surface map showing atmospheric pressure.

Coriolis Effect

Pressure gradient is the driving force of winds and directs winds perpendicular to isobars. But winds do not generally move perpendicular to isobars because of the Coriolis effect. **Coriolis effect** is an apparent deflection of winds caused by the Earth's rotation. *Every free-moving object, including winds, deflects to the right of its path of motion in the northern hemisphere, and to the left of its path of motion in the southern hemisphere.* The movement of air near the Earth's surface is at an angle across the isobars toward the area of low pressure.

Isobars on weather maps are closed circles, indicating areas of high and low pressure. Low pressure cells are called **cyclones** and show where air is rising. High pressure cells are **anticyclones**, showing where air is sinking. In the northern hemisphere, winds circulate counter-clockwise as they move toward the center of low pressure (cyclones). In the northern hemisphere, winds blow clockwise around high pressure cells (anticyclones). In the southern hemisphere, Coriolis effect deflects winds to the left (opposite direction from the northern hemisphere), so circulation is clockwise around cyclones and counter-clockwise around anticyclones.

Figure 5.3 is a simplified map of barometric pressure at the Earth's surface *in the northern hemisphere.*

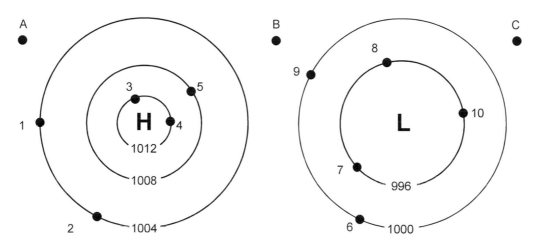

Figure 5.3.   Northern hemisphere surface pressure map.

1. On this figure draw short straight arrows to indicate wind direction for locations 1–10.

2. The wind would blow *from* what direction at:

                    Point A: _____

                    Point B: _____

                    Point C: _____

Drawing Isobars

Barometric pressure shown by isobars is one of the most important features of a weather map. Drawing isobars will help you to determine how surface winds move over large areas.

On Figure 5.4 is a map showing barometric pressure (mb) observed simultaneously at many National Weather Service stations. Draw isobars for the entire map using an interval of 4 mb: 992, 996, 1000, 1004, 1008, etc. Label the isobars and the center of high pressure (H) and low pressure (L). Be sure that isobars do not cross. Use a pencil and draw lightly until you are sure the isobar positions are correct, then darken the lines. Finally, draw short straight arrows across the isobars to show the directions of surface winds.

Answer the following questions based on the barometric pressure map you have created.

1. What is the highest pressure on this map and where is it located?

    highest pressure: _____ mb

    location: _____

2. What is the lowest pressure on this map and where is it located?

    lowest pressure: _____ mb

    location: _____

3. In what direction is the wind blowing *from*

    at your location: _____

    _____ : _____

    _____ : _____

    _____ : _____

A general rule about the relationship between barometric pressure and surface wind is: *Stand with the wind at your back with your arms outstretched to either side, and the lower pressure will be towards your left.* This is called **Ballot's law**. Although not really a law of science, it works quite well with two provisions. First, you must be in the northern hemisphere. Second, your left arm should be aimed halfway between straight left and straight ahead.

Use this technique to determine the position of high and low pressure relative to your position. Measure the following information in an area away from buildings or topographic features that could obstruct winds:

    wind direction: _____

    wind speed: _____ km/hr

    barometric pressure: _____ mb

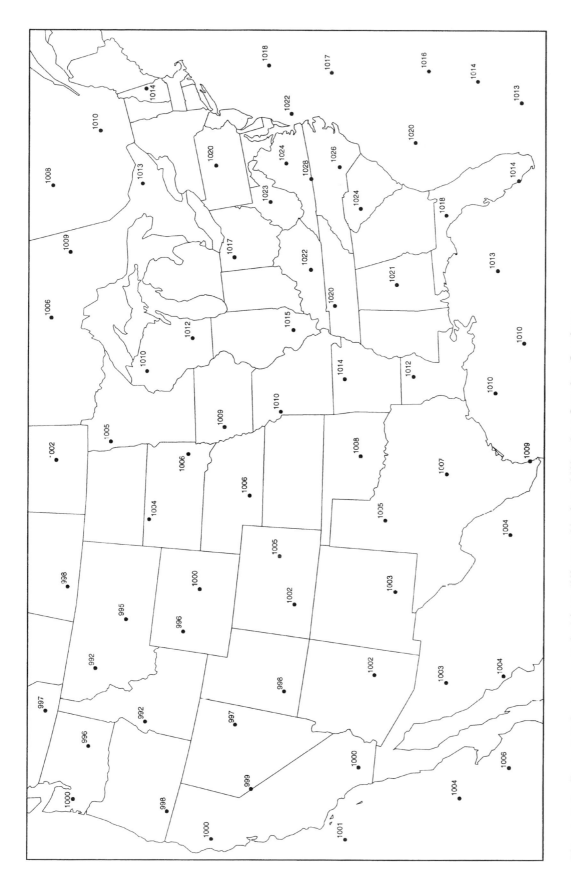

Figure 5.4   Barometric pressure (mb) at different National Weather Service Stations.

49

Figure 5.5 is a map of the United States. On this map (1) write the barometric pressure reading you have recorded at your location and (2) draw a short straight arrow indicating wind direction. Taking into account how surface winds circulate around cyclones (as shown in Figure 5.3 and 5.4) indicate the likely position of the center of low pressure (L). Draw isobars around the Low. Remember that surface winds move at an angle across isobars toward the center of low pressure. Also, include the possible position for high pressure (H) and draw additional isobars. You have already shown wind direction at your location. Draw a few short straight arrows across the isobars to show the directions of surface winds at other locations.

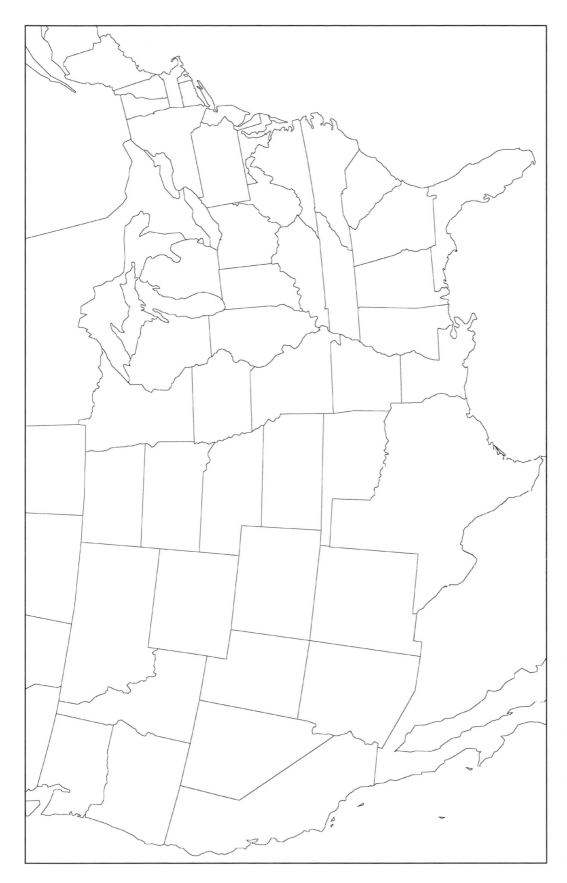

Figure 5.5 Map of the United States.

Atmospheric Pressure and Winds: Internet Resources

Surface Pressure and Winds

http://weather.colorado.edu/pages/mainnat.html This site, maintained by the University of Colorado, has many weather images. The ***Composite*** image is an infrared satellite photograph showing clouds and surface pressure. Note that the times given on this site are Mountain Standard Time or Mountain Daylight Savings Time.

http://www.ems.psu.edu/wx/usstats This site, maintained by the Pennsylvania State University Department of Meteorology, provides the same information on surface pressure and winds as the above site. It deals mostly with the northeastern U.S. However, ***Hourly U.S. Weather Statistics*** are available and includes graphical displays of surface temperature, winds, pressure, dew point, and cloud cover for the past 10 days.

Date: _____ Time: _____

Surface pressure at your location?

Lat. _____ ; Long. _____

Currently: _____ mb

24 hours ago: _____ mb

48 hours ago: _____ mb

What is the current wind speed at your location? _____

Note the proximity of high and low pressure centers relative to your location.

How have these pressure cells moved during the past 48 hours? Note the direction and speed they are moving.

http://www.nws.noaa.gov/ The National Weather Service Homepage has regularly updated weather information and forecasts for the U.S. and throughout the world. Select ***Weather Maps*** to access ***Surface Analysis of North America***. You can print a map showing surface pressure and winds.

LAB EXERCISE
6

Atmospheric Humidity

Equipment

psychrometer

Water vapor is a small proportion of the atmosphere, ranging from about 1–3 percent in most regions of the world. However, its abundance is one of the most important atmospheric conditions determining weather and influencing regional climates. In this exercise you will measure relative humidity and address problems that deal with the relationship between temperature, water vapor capacity, relative humidity, and condensation.

Water at the earth's surface or in the lower atmosphere occurs as liquid, solid, or gas depending primarily on temperature. Water vapor, an invisible gas, enters the atmosphere by evaporation. The amount (or density) of water vapor is referred to as **specific humidity**, and is usually expressed as grams per kilogram.

The amount of water vapor the air can hold depends on air temperature. As temperature increases, the air can hold more water. As temperature decreases, its ability to hold moisture decreases. The maximum amount of water vapor it can hold at any given temperature is referred to as **capacity**. The ratio between the **specific humidity**, the actual amount of water in the air, and **capacity** (how much it can hold at that temperature), is called **relative humidity** (RH), which is expressed as a percentage:

$$RH = (SH/CAP) * 100$$

where: SH = specific humidity
CAP = capacity

As the air temperature changes, the air's capacity to hold water and relative humidity also changes. For example, on the first line of Table 6.1 shown below, a parcel of air has a temperature of 25°C. Capacity at this temperature (as indicated in Figure 6.1) is 20 g/kg. Assume the specific humidity is 5 g/kg (although it could be higher or lower). Since the specific humidity (SH) is ¼ of capacity, the relative humidity (RH) is 25 percent. In other words, the air has ¼ of the amount of water it is capable of holding.

Now assume that the specific humidity (the amount of water in the air) does not change, but the temperature decreases. With lower temperature, the air is capable of holding less water. So, while the amount of water in the air remains the same, the capacity is reduced, and therefore the relative humidity increases. On the second line of the table shown below the air temperature decreased from 25°C to 20°C. The capacity (CAP) decreased (15g/kg) so the

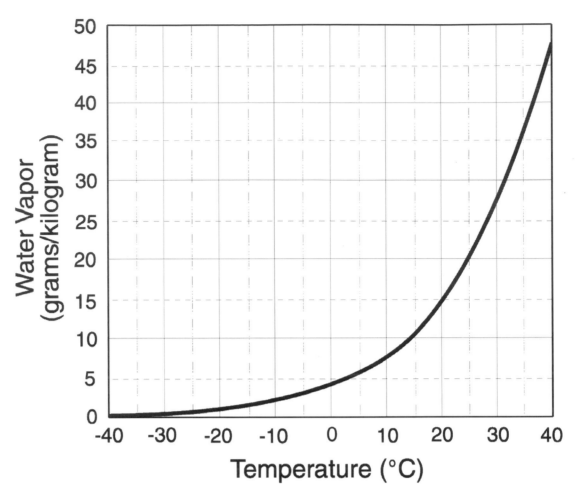

Figure 6.1 **Saturation curve showing the maximum amount of water vapor (capacity) that can be held at a given temperature.**

relative humidity (SH/CAP) increased. As the air temperature continues to decrease, the relative humidity increases.

Example Table 6.1 Determine Relative Humidity of the Air at Various Temperatures

Temp(°C)	CAP	SH	RH(%)
	g/kg		
25	20	5	25
20	15	5	33
15	10	5	50
10	8	5	63
5	5	5	100

For the following questions refer to Figure 6.1.

1. What is the capacity of a parcel of air that has a temperature of 20°C?

_____ grams per kilogram

2. What is the capacity of a parcel of air that has a temperature of 10°C?

_____ grams per kilogram

3. If a parcel of air is 10°C and the specific humidity is 5 grams of water per kilogram, what is the relative humidity?

_____ %

Fill in the blanks on the following table. Use Figure 6.1 to determine the capacity for the corresponding air temperatures.

Temp(°C)	Capacity	Specific Humidity	Relative Humidity
		g/kg	%
30	28	10	_____
20	_____	10	_____
15	_____	10	_____
10	_____	5	_____
_____	_____	_____	_____
_____	_____	_____	_____

4. Minimum daily temperature usually occurs at sunrise. Typically, the temperature rises until reaching a maximum during the mid-afternoon and then slowly declines. Based on these daily temperature changes, how would the relative humidity change during the day?

Dew point is the temperature at which the air must be cooled to reach a relative humidity of 100 percent. At a relative humidity of 100 percent the air is saturated. In Table 6.1, since the specific humidity (SH) is 5 g/kg, the air had to be cooled so that the capacity (CAP) equaled the same.

For the following questions refer to Figure 6.1.

1. What is the dew point if the specific humidity is 15g/kg?

_____ °C

57

2.   What is the dew point if the specific humidity is 25g/kg?

_____ °C

3.   If the air temperature is 10°C and the relative humidity is 50 percent, what is the dew point? (Hint: First, determine the specific humidity.)

_____ °C

4.   If the air temperature is 15°C and the relative humidity is 33 percent, what is the dew point?

_____ °C

You can easily determine the relative humidity using a sling psychrometer, which has two thermometers. One thermometer (the dry-bulb) measures the air temperature. The other (the wet-bulb) has a cloth wick at the base which is wetted. Saturate the wick (the wet-bulb) and then swing the psychrometer for 20 seconds. Read the temperature of the wet bulb thermometer and then repeat this process until the wet-bulb temperatures do not change between readings. Record the final wet-bulb and dry-bulb temperatures. The difference between these readings is the **wet-bulb depression**.

The amount of evaporation from the wet bulb indicates the amount of water vapor that is in the air. If the air is near saturation little water will evaporate from the wet bulb, and therefore the wet bulb depression, or difference in temperature between the two thermometers will be small. In contrast, if the air is very dry, water will rapidly be lost from the wet bulb, and the wet bulb depression will be high.

The temperature and, therefore, humidity readings will vary on different surfaces. Vegetated surfaces and those that are shaded will typically have cooler temperatures compared to paved surfaces and those in direct sunlight. Take readings from the sling psychrometer over the different surfaces listed below and fill in the following table. If you are using Celsius temperatures, refer to Table 6.2 to determine the relative humidity. Use Table 6.3 if you are using Fahrenheit.

Caution! Sling psychrometers will easily break. Please handle them carefully and make sure they do not strike other objects.

	Black Asphalt	Vegetated Surface Full Sun	Shade	
Wet Bulb	_____	_____	_____	_____
Dry Bulb	_____	_____	_____	_____
Wet Bulb Depression	_____	_____	_____	_____
Relative Humidity (%)	_____	_____	_____	_____
Dew Point	_____	_____	_____	_____

1. Over which surface was the air temperature highest? Why?

2. Which surfaces had the highest relative humidity? Why?

3. Would you expect air temperature and relative humidity to be significantly different on a cloudy day? How would these conditions vary?

Table 6.2 Percent Relative Humidity, Temperature in °Celsius

Air Temperature	Wet Bulb Depression														
	1	2	3	4	5	6	7	8	9	10	11	12	13	14	15
-5	76	52	29	7											
-4	77	55	33	12											
-3	78	57	37	17											
-2	79	60	40	22											
-1	81	62	43	26	8										
0	81	64	46	29	13										
1	83	66	49	33	17										
2	84	68	52	37	22	7									
3	84	70	55	40	26	12									
4	85	71	57	43	29	16									
5	86	72	58	45	33	20	7								
6	86	73	60	48	35	24	11								
7	87	74	62	50	38	26	15								
8	87	75	63	51	40	29	19	8							
9	88	76	64	53	42	32	22	12							
10	88	77	66	55	44	34	24	15	6						
11	89	78	67	56	46	36	27	18	9						
12	89	78	68	58	48	39	29	21	12						
13	89	79	69	59	50	41	32	23	15						
14	90	79	70	60	51	42	34	26	18	8					
15	90	80	71	61	53	44	36	27	20	13					
16	90	81	71	63	54	46	38	30	23	15	8				
17	90	81	72	64	55	47	40	32	25	18	11				
18	91	82	73	65	57	49	41	34	27	20	14				
19	91	82	74	65	58	50	43	36	29	22	16	10			
20	91	83	74	66	59	51	44	37	31	24	18	12	6		
21	91	83	75	67	60	53	46	39	32	26	20	14	9		
22	92	83	76	68	61	54	47	40	34	28	22	17	11	6	
23	92	84	76	69	62	55	48	42	36	30	24	19	13	8	
24	92	84	77	69	62	56	49	43	37	31	26	20	15	10	5
25	92	84	77	70	63	57	50	44	39	33	28	22	17	12	8
26	92	85	78	71	64	58	51	46	40	34	29	24	19	14	10
27	92	85	78	71	65	58	52	47	41	36	31	26	21	16	12
28	93	85	78	72	65	59	53	48	42	37	32	27	22	18	13
29	93	86	79	72	66	60	54	49	43	38	33	28	24	19	15
30	93	86	79	73	67	61	55	50	44	39	35	30	25	21	17
31	93	86	80	73	67	61	56	51	45	40	36	31	27	22	18
32	93	86	80	73	67	61	56	51	45	41	37	32	28	24	20
33	93	86	80	74	68	63	57	52	47	42	38	33	29	25	21
34	93	87	81	75	69	63	58	53	48	43	39	35	30	26	23

Table 6.3 Percent Relative Humidity, Temperature in °Fahrenheit

Air Temperature | **Wet Bulb Depression**

Air Temp	1	2	4	6	8	10	12	14	16	18	20	22	24	26	28
24	84	73	47	22											
26	87	75	51	27	4										
28	88	76	54	32	10										
30	89	78	56	36	16										
32	89	79	59	39	20	2									
34	90	81	62	43	25	8									
36	91	82	64	46	29	13									
38	91	83	66	50	33	17	3								
40	92	83	68	52	37	22	7								
42	92	85	69	55	40	26	12								
44	93	85	71	56	43	30	16	4							
46	93	86	72	58	45	32	20	8							
48	93	86	73	60	47	35	23	12	1						
50	93	87	74	61	49	38	27	16	5						
52	94	87	75	63	51	40	29	19	9						
54	94	88	76	64	53	42	32	22	12	3					
56	94	88	75	65	55	44	34	25	16	7					
58	94	88	77	66	56	46	37	27	18	10	1				
60	94	89	78	68	58	48	39	30	21	13	5				
62	94	89	79	69	59	50	41	32	24	16	8				
64	95	90	79	70	60	51	43	34	26	18	11	1			
66	95	90	80	71	61	53	44	36	29	21	14	5			
68	95	90	80	71	62	54	46	38	31	23	16	7			
70	95	90	81	72	64	55	48	40	33	25	19	11	2		
72	95	91	82	73	65	57	49	42	34	28	21	14	5		
74	95	91	82	74	65	58	50	43	36	29	23	19	14	7	
76	96	91	82	74	66	59	51	44	38	31	25	21	16	11	
78	96	91	83	75	67	60	52	46	39	33	27	22	17	12	5
80	96	91	83	75	68	61	54	47	41	35	29	23	18	12	7
82	96	92	84	76	69	61	55	48	42	36	30	25	20	14	10
84	96	92	84	76	69	62	56	49	43	37	32	26	21	16	12
86	96	92	84	77	70	63	57	50	44	39	33	28	23	18	14
88	96	92	85	77	70	64	57	51	46	40	35	30	25	20	15
90	96	92	85	78	71	65	58	52	47	41	36	31	26	22	17
92	96	92	85	78	72	65	59	53	48	42	37	32	28	23	19
94	96	93	85	79	72	66	60	54	49	43	38	33	29	24	20
96	96	93	86	79	73	66	61	55	50	44	39	35	30	26	22
98	96	93	86	79	73	67	61	56	50	45	40	36	32	27	23

Atmospheric Humidity: Internet Resources

Humidity

http://www.ems.psu.edu/wx//index.html This site, maintained by the Pennsylvania State University Department of Meteorology, has weather maps and data. It deals mostly with the northeastern U.S. However, *Hourly U.S. Weather Statistics* are available and includes graphical displays of surface temperature, winds, pressure, dew point, and cloud cover for the past 10 days.

Select *Dewpoint*, which shows a current dew point map for the U.S.

What is:

highest dew point in the U.S.? _____ location: _____

lowest dew point in the U.S.? _____ location: _____

at your location? Lat. _____ ; Long. _____

   today? _____

   one day ago? _____

   two days ago? _____

   one week ago? _____

Select *Cloud Cover*.

What areas of the U.S. have a heavy cloud cover?

Is there a relationship between dew point and cloud cover? If yes, describe this relationship.

Weather Satellite Images

http://www.people.fas.harvard.edu/~dbaron/sat/ This site has an extensive collection of weather satellite images showing cloud cover. Select *GOES-EAST* or *GOES-WEST* for current images of North America.

http://www.cira.colostate.edu/ This site is from the Cooperative Institute for Research in the Atmosphere at Colorado State University. You can obtain *Satellite Images* from geostationary and polar orbiting satellites. These images are updated several times each day.

Adiabatic Processes and Precipitation

Equipment

psychrometer

The rising of air is the most common cooling mechanism in the atmosphere and is the direct cause of most condensation and precipitation. The purpose of this exercise is to develop an understanding of the relationships between the rising of air, changes in relative humidity, and the elevation or level of condensation and cloud formation.

If an air mass rises, it will expand because of decreasing atmospheric pressure. As the air expands there are fewer molecular collisions that generate heat and therefore the air will cool. Rising air may cool to its dew point and clouds will form. If additional rising occurs after the dew point is reached, precipitation will likely occur. Conversely, if the air descends in altitude it is compressed and will heat, therefore, increasing its ability to hold moisture. Cloud development and precipitation will not occur if air is sinking.

The change in temperature caused by air rising or sinking is called **adiabatic cooling and heating**. It is important to make a distinction between the adiabatic process and the increase in air temperature with altitude (in an air mass that is not rising or sinking) which is referred to as the environmental lapse rate (ELR).

If air rises but has not cooled to its dew point, the air will cool at a constant rate of 10°C/ 1 km (1°C/100m). This is called the **dry adiabatic rate** (DAR). If the air rises enough so that it cools to the dew point and condensation occurs, the air will continue to cool but at a lower rate of about 6°C/1 km (0.6°C/100m). This is known as the **saturated adiabatic rate (SAR)**. When air begins to descend, the temperature will heat at the dry adiabatic rate (10°C/1 km) and immediately exceed the dew point.

Orographic Precipitation

Orographic precipitation results from air cooling as it rises over mountains or topographic barriers. As it rises, the air cools at the dry adiabatic rate (10°C/1 km or 1°C/100m). As the air sinks on the leeward side, it warms at the same rate.

1. Figure 7.1 shows an air mass that is forced to rise over a mountain that is 4000m elevation. The air cools as it rises. But in this problem, assume that the air does not cool enough to reach the dew point, so no condensation occurs. Because the relative humidity is less than 100 percent, use the DAR to determine the rate of temperature change. The temperature at 1000m elevation on the windward side of the mountain is 30°C. Calculate and indicate the temperature for the elevations shown on Figure 7.1.

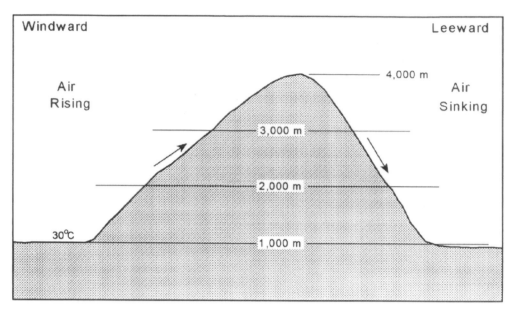

Figure 7.1   Calculate temperature for various elevations.

If your calculations are correct, temperature should be the same on both the windward and leeward sides of the mountain at the same elevation.

2.   Figure 7.2 shows an air mass that is forced to rise over a mountain at another location. The mountain summit is 4000m. The temperature at the foot of the mountain (1000m elevation) is the same as in the previous problem. But in this case assume that the humidity is higher so that condensation occurs at 2000m as the air ascends and cools on the windward side.

   Remember, use the dry adiabatic rate if the relative humidity is less than 100 percent. Use the saturated adiabatic rate only if air continues to rise after the dew point is reached.

   Calculate and indicate the temperature for the elevations shown on Figure 7.2.

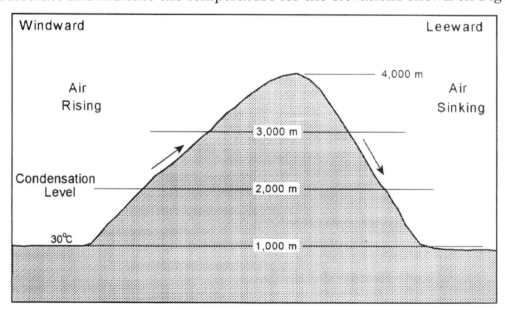

Figure 7.2   Calculate temperature for various elevations.

After reaching the summit the air begins to descend and is adiabatically heated. The warmer air has a higher water vapor capacity than at the summit and therefore the relative humidity immediately drops below 100 percent.

a. If your calculations are correct, air temperature at the same elevation on the windward and leeward sides of the mountain are different. Why?

b. The relative humidity is usually lower on the leeward side of mountains. What are the two factors that cause drier conditions there?

(1) _____

(2) _____

c. A **rainshadow** refers to the dry conditions on the leeward side of mountains. This is one of the major causes of desert regions in the world. Identify a rainshadow region in North America and the mountain range(s) causing this effect.

Convection

The ascent of air is not always forced as demonstrated in the previous problems. **Convection** refers to air that rises spontaneously or freely. The earth's surface heats unevenly and some surfaces have a higher temperature than others at the same time. Since the earth's surface is the major heat source for the lower atmosphere, air temperature near the ground will vary from one place to another.

If a parcel of air is warmer than the surrounding air, it will tend to rise, cooling at the dry adiabatic rate. Even though the air is cooling as it rises, as long as it remains warmer than the surrounding air it will continue to rise. In most cases, the air cools and eventually sinks back toward the surface before reaching the dew point, and therefore, no condensation occurs.

However, if a parcel of air rises high enough to reach the dew point, clouds will form.

In the following problems, use the DAR to calculate the temperature of a rising parcel of air, and the Environmental Lapse Rate (ELR) to calculate the temperature of the surrounding air.

Environmental Lapse Rate (ELR) = 7°C/1000m
(for the surrounding air that is not rising)

Dry Adiabatic Rate (DAR) = 10°C/1000m

As shown in Chapter 6, the dew point of a parcel of air can easily be calculated at the surface. However, the dew point changes as air rises. Remember that absolute humidity is a measure of the weight of water vapor per volume of air (g/m^3). As air rises it expands. Because of the change in volume, as air rises the dew point decreases. To accurately determine the level of condensation, you must take into account the **dew point lapse rate (DPLR)**, which is a decrease in the dew point of 2°C/1000m (.2°C/100m).

1. In the following problem, the temperature of a parcel of air at the ground is 30°C, which is 6°C warmer than the surrounding air (24°C). The dew point is 6°C. The warm parcel of air begins to rise and cool. Enter temperature of the rising parcel of air and the surrounding air, and dew point for each elevation on the table.
    The parcel of air will rise only if it is warmer than the surrounding air. At the elevation where the dew point equals the temperature of the rising air, the relative humidity is 100 percent.

Elevation (m)	Temperature(°C)		Dew Point(°C)
	ELR	DAR	
4000-	_____	_____	_____
3500-	_____	_____	_____
3000-	_____	_____	_____
2500-	_____	_____	_____
2000-	_____	_____	_____
1500-	_____	_____	_____
1000-	_____	_____	_____
500-	_____	_____	_____
surface-	24	30	6
	↑	↑	
	surrounding air	rising air	

In this problem, the relative humidity (and therefore dew point) at the surface was low. If your calculations are correct, the column of air stopped rising before reaching the dew point, and no condensation or precipitation occurred.

a. How high (or to what elevation) did the air parcel rise?

    _____ m

b. How much higher did the parcel of air need to rise to reach the dew point?

    _____ m

2. In the following problem, the temperature at the surface is 30°C, which is the same as in the previous problem. The only difference is that humidity (and dew point) at the surface is higher. Higher humidity increases the probability that the rising air will reach the condensation level.

   Calculate the temperature of a rising parcel of air. As in the previous problem, the air will only rise if it is warmer than the surrounding air.

   Calculate and enter temperature and dew point on the following table. Remember to use the saturated adiabatic rate if the relative humidity is 100 percent.

| Elevation (m) | Temperature(°C) | | Dew Point(°C) |
	ELR	DAR	
4000-	_____	_____	_____
3500-	_____	_____	_____
3000-	_____	_____	_____
2500-	_____	_____	_____
2000-	_____	_____	_____
1500-	_____	_____	_____
1000-	_____	_____	_____
500-	_____	_____	_____
surface-	24	30	18
	↑	↑	
	surrounding air	rising air	

a. At what elevation was dew point reached?

   _____ m

b. What is the difference in temperature between the rising column of air and the surrounding air at:

   2000m? _____

   3000m? _____

   4000m? _____

69

c. Note that after reaching the condensation level, the temperature of the rising air becomes progressively warmer than the surrounding air. Why is the temperature difference greater as the air rises (what is the heat source)?

d. Based on your findings in the last two problems, why is convectional precipitation much more common in humid as opposed to desert regions?

3. In this problem you will measure temperature and humidity immediately outside your building to determine the level of condensation, or the cloud base.
Use a sling psychrometer over a vegetated surface that is in full sunlight to determine:

dry bulb temperature _____ °C

wet bulb temperature _____ °C

wet bulb depression _____ °C

Use Table 6.2 to determine relative humidity: _____ %

Use Figure 6.1 to determine the dew point: _____ °C

Enter the temperature and dew point at the surface on the following table. Calculate and enter these values for the elevations given.

Elevation(m)	Temperature(°C)	Dew Point(°C)
5000-	_____	_____
4500-	_____	_____
4000-	_____	_____
3500-	_____	_____
3000-	_____	_____
2500-	_____	_____
2000-	_____	_____
1500-	_____	_____
1000-	_____	_____
500-	_____	_____
surface-	_____	_____

a.   What is the elevation of the cloud base?

_____ m

b.   What is the dew point at the level of condensation?

_____ °C

On Figure 7.3 plot a line showing the temperature decrease with increasing elevation. Remember to use the DAR if the relative humidity is less than 100 percent, and the SAR once the air has become saturated. Also, draw a line showing the decrease in dew point with increasing elevation (2°C/1000m). The elevation where the temperature and dew point lines meet indicates the elevation of the cloud base (relative humidity is 100 percent). Label all lines and mark the level of condensation.

Figure 7.3   Temperature decrease with increasing elevation.

72

Air Masses, Cyclonic Storms, and Fronts

Air masses and pressure systems cover extensive areas. Viewing atmospheric conditions over a large geographic region is necessary to understand weather at any single position. The objective of this exercise is for you to (1) become familiar with surface weather maps, and (2) to analyze weather systems and make short-term weather forecasts based on these maps.

The National Weather Service collects surface weather data from dozens of weather recording stations around the country. Every day new data are plotted on maps to show large scale atmospheric phenomena such as air masses, cyclonic storms, fronts, winds, cloud cover, and precipitation. The location of each recording station is indicated by a circle on the map. The weather station model (Figure 8.1) refers to symbols around this circle indicating weather conditions at the time of data collection. Most weather stations do not report all of the information shown on this sample. But air temperature, dew point temperature, and barometric pressure are almost always included. Note that pressure is indicated by the last three digits of millibar readings. A 9 or 10 must be placed in front of the 3 digits. For example, a pressure reading of 1024.6 mb would be shown as 246.

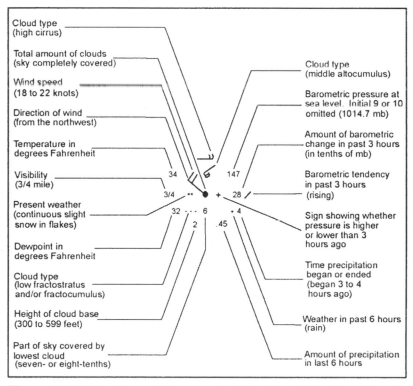

Figure 8.1   Weather station model. (National Weather Service.)

The amount of cloud cover is represented by the type of shading inside the circle (Figure 8.2). Wind direction is shown by the wind arrow, or line that indicates the direction of wind toward that station. The tails or feathers on the end of the wind arrow indicate wind speed. The weather map symbols used on these maps are shown below in Figure 8.2.

Front symbols:

▲▲▲▲ Cold Front   ●●●● Warm Front

▲▲▲▲ Occluded Front   ▼▲▼▲ Stationary Front

Total sky cover:

○ No clouds   ◔ Six-tenths

◍ Less than one-tenth or one-tenth   ◕ Seven-tenths or eight-tenths

◷ Two-tenths or three-tenths   ◑ Nine-tenths or overcast with openings

◔ Four-tenths   ● Completely overcast

◑ Five-tenths   ⊗ Sky obscured

Wind entries:

◎ calm

— 1-2 knots

∟ 3-7

∟ 8-12

∟ 13-17

∟ 18-22

∟ 23-27 knots

∟ 28-32

∟ 33-37

∟ 38-42

∟ 43-47

∟ 48-52

∟ 53-57 knots

∟ 58-62

∟ 63-67

∟ 68-72

∟ 73-77

∟ 103-107

Figure 8.2   Weather station symbols.

1. Figure 8.3 is an example of a weather map showing atmospheric conditions on February 11, 1988. An area of low pressure is centered over Kentucky (Lat. 37°N, Long. 86°W). Counter-clockwise circulation around the cyclonic storm has moved cold air unusually far to the south. Note the air mass boundaries indicating the edge of the cold air mass. These are shown by a cold front extending from Kentucky to northern Mexico and a stationary front extending from Canada south to New Mexico. The shaded areas on the map indicate areas of precipitation.
    a. What are the highest and lowest barometric pressure readings on the map and where are they located?

        high pressure: _____

        location: _____

74

Figure 8.3 Surface weather map for February 11, 1988.

75

low pressure: _____

location: _____

b. What is the coldest temperature on the map and where is it located?

low temperature: _____

location: _____

c. What is the wind direction and speed in Oklahoma or northern Texas?

d. Compare the temperature at San Francisco to stations in the following states that are at about the same latitude (35°N):

Nevada

Oklahoma

North Carolina

e. Why is winter temperature at the California coast warmer than most other areas in the country?

f. Referring to the weather station closest to you, give the following information:

Temperature: _____

Wind direction: _____

Wind speed: _____

Barometric pressure: _____

2. Two weather maps show Hurricane Andrew in the Gulf of Mexico on August 24, 1992 and two days later, August 26, 1992 (Figures 8.4 and 8.5). Andrew crossed southern Florida causing extensive damage, and less than two days later hit the coast of Louisiana. Note that Andrew moved generally east to west following the trade winds. In the central U.S. a cold front is moving in the opposite direction because it is at a high latitude and therefore in the zone of the westerlies.

    a. What is the latitude and longitude of Andrew on:

        August 24? _____

        August 26? _____

    b. What is the average speed that the center of the hurricane is moving between August 24 and 26?

        _____

    c. In which direction do winds circulate around the hurricane?

        _____

    d. What is the barometric pressure in the center of Andrew on August 24?

        _____

    e. On August 26, the highest barometric pressure is

        _____ mb

        location: _____

    f. Andrew moved inland and then dissipated. Why do hurricanes lose power over land?

    g. Describe the change of the position of the cold front in the central U.S. between August 24 and 26.

    h. Generally, what are the differences in air temperature on either side of the cold front?

Figure 8.4 Surface weather map for August 24, 1992.

Figure 8.5   Surface weather map for August 26, 1992.

3. By looking at daily weather maps over a period of several days, you can easily track the movement of atmospheric phenomena, such as cyclonic storms, high pressure, and fronts. A series of four weather maps (Figures 8.6–8.9) covering the period of October 16–19, 1990 shows changing meteorological conditions through time. The map of October 16 shows low pressure in the western U.S. and high pressure located in the eastern U.S. The cyclonic storm (low pressure) strengthens and begins to move to the east. The movement of the storm is indicated by a chain of arrows. Associated with the storm are well defined cold and warm fronts.

    a. What is the average speed that the center of the cyclonic storm was moving between October 16–19?

    _____

    b. Describe the change in position of the cold front during this 4-day period.

    c. What was the general wind direction and speed in the region of the Ohio River valley on October 16 and 18?

    d. The center of the cyclonic storm passed over Lake Michigan. Use the isobars to determine the changes in barometric pressure during this four day period.

        October 16: _____ mb

        October 17: _____ mb

        October 18: _____ mb

        October 19: _____ mb

    e. What is the relationship between areas of precipitation and the position of high pressure, low pressure, and fronts?

Figure 8.6 Surface weather map for October 16, 1990.

Figure 8.7　Surface weather map for October 17, 1990.

Figure 8.8 Surface weather map for October 18, 1990.

83

Figure 8.9 Surface weather map for October 19, 1990.

f.   Referring to the weather station closest to you:

    (1)   Give the following information for October 16 (Figure 8.6):

            Temperature: _____

            Wind direction: _____

            Wind speed: _____

            Barometric pressure: _____

    (2)   Refer to the maps showing conditions during the next three days, October 17–19 (Figures 8.7–8.9). Describe how the conditions listed above changed during this period. Also describe the relationships between these changes and the movement of the cyclonic storm and high pressure, and the passage of fronts.

Air Masses, Cyclonic Storms, and Fronts: Internet Resources

U.S. Weather Maps

http://nws.noaa.gov/ This site is from the National Weather Service. From this site you can look at current weather conditions and forecasts for the U.S. and international locations. It also has storm predictions, marine weather forecasts, and updated satellite imagery. Select **weather maps**. There are maps updated every day showing surface conditions and barometric pressure at different elevations.

Select **Surface Analysis** for **North America**. Based on this map, describe current weather conditions at your location:

Lat. _____ ; Long. _____

Date: _____ Time: _____

temperature: _____

dewpoint: _____

surface pressure: _____

wind direction and speed: _____

Where are the nearest fronts located and in what direction are they moving?

What is the proximity of high and low pressure to your location?

http://weather.colorado.edu/pages/mainnat.html This site, maintained by the University of Colorado, has many weather images. The **Composite** image is an infrared satellite photograph showing clouds, surface pressure, and fronts.

Based on the composite image, what is the relationship between surface pressure and cloud cover?

Atlantic Hurricane Tracking, Maps, and Images

http://fermi.jhuapl.edu/hurr/index.html This site includes maps and satellite images for hurricanes in the Atlantic. The names for all Atlantic northern hemisphere hurricanes are listed for the past few years. It is maintained by the John Hopkins University Applied Physics Laboratory

Select **Track Maps for Selected Earlier Atlantic Hurricanes**. Here are track maps showing wind speed for some of the earlier and more powerful hurricanes, including Gilbert (1988), Hugo (1989), and Andrew (1992). Earlier in this chapter are two weather maps showing Hurricane Andrew. Select **Andrew** and note that wind speed decreased immediately after Andrew moved inland. You will see a similar pattern for **Hugo** and **Gilbert**.

What is the general direction hurricanes move in (1) the Caribbean Sea and (2) a region farther north in the Atlantic? Relate the direction they move to the global circulation model.

Why do hurricanes lose intensity when they move inland?

Note to lab instructors: Hurricane tracking maps for the Atlantic and Pacific Oceans can be downloaded from both of the above sites. These may be useful for students to track storms if the course occurs during the hurricane season.

http://www.ncdc.noaa.gov/ol/climate/severeweather/hurricanes.html This NOAA site shows where hurricanes from 1899 to 1996 moved inland.

Based on the maps showing landfalling hurricanes, what shoreline areas of the eastern U.S. are most affected by hurricanes?

http://www.nhc.noaa.gov/ The National Hurricane Center tracks and analyzes tropical storms, and issues forecasts and warnings. This site provides the latest satellite images and information on active tropical storms. This site also has **Storm Archives**. You can select years and names of storms for specific information.

http://www.weather.unisys.com/hurricane/atlantic/index.html This site has a list of hurricanes since 1886. You can select the year and then access detailed information on any northern hemisphere Atlantic hurricane.

http://photolib.noaa.gov/ This site has a collection of photographs from the National Oceanic and Atmospheric Administration. You can select **Search** and then enter hurricanes, tornadoes, lightening, fronts, etc. for many weather photographs.

Weather Analysis

In this exercise weather data at your location will be collected for a period of several days and then analyzed. The purpose of the exercise is to determine how major meteorological events can affect atmospheric conditions and to examine interrelationships among climatic parameters. For this exercise you are required to collect weather data for a period of up to 2 weeks before analysis begins.

Data Collection

Weather data will be collected once each day. The data will be collected at the same time every day. There are several sources that can be used, including radio or television weather reports. The data collection must be consistently from the same source.

The data will be recorded on the weather data log on the next page. If you miss collecting some data, do not use other sources or get the information from someone else. Omissions should be avoided. But missing minor subsets will not be critical. Data inconsistencies, caused by using different sources, may affect or invalidate your analysis.

Also, you should attempt to identify major changes in atmospheric conditions or meteorological events that occurred during the period of data collection, such as change in the positions of high pressure systems and cyclonic storms relative to your position, the passage of fronts, convectional precipitation, temperature inversions, etc. Record these conditions with the time and day they occurred.

Name _____

Group _____

Recording
Time _____

Weather Data Log

Date	Temp °C	Wind Direction	Wind Speed	Relative Humidity	Barometric Pressure	Precip (cm)	Cloud Cover (%)

After completing the above log, plot points for temperature and humidity on Figure 9.1, and barometric pressure and precipitation on Figure 9.2 for each day. Connect the points with lines for the different atmospheric variables shown. The lines will help you to visually interpret the change in conditions over time.

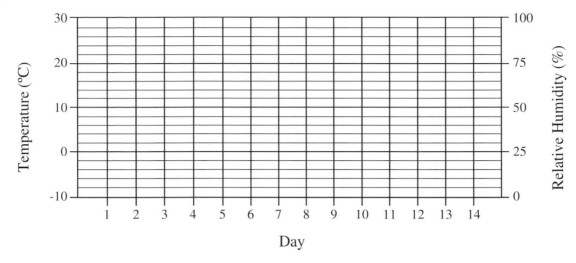

Figure 9.1   Temperature and humidity.

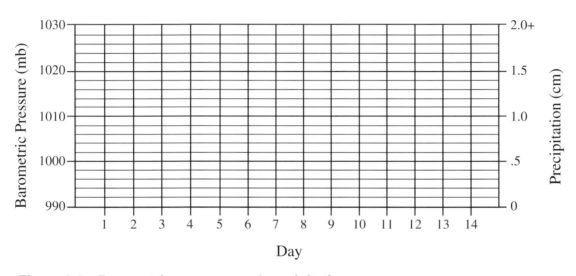

Figure 9.2   Barometric pressure and precipitation.

Use your Weather Data Log and the above graphs to answer the following questions.

1.   What is the temperature range (max.–min.)? _____

   maximum temperature: _____

   minimum temperature: _____

2.   What is the relative humidity range (max.–min.)? _____

   maximum RH: _____

   minimum RH: _____

3. Describe the relationship between temperature and humidity shown on Figure 9.1. Is there a direct or inverse relationship, and is this relationship strong or weak?

4. What is the barometric pressure range (max.–min.)? _____

   maximum pressure: _____

   minimum pressure: _____

5. It would not be unusual for no precipitation to occur over such a short period of time. Did precipitation occur during the period of data collection, and if so, describe any relationships between barometric pressure and precipitation (Figure 9.2).

Listed below are atmospheric phenomena that may or may not have affected your data. Compare the data that you have recorded in the Weather Data Log to the occurrence of these events (either that you noted or that is provided by the lab instructor). Describe and analyze these relationships. Include in these discussions when appropriate: changes in wind direction and speed, cloud cover and precipitation, temperature, humidity, and pressure.

1. Change in position of cyclonic storms:

2. Change in position of high pressure:

3. Passage of fronts:

4. _____

5. _____

Weather Analysis: Internet Resources

Climate Prediction

http://www.cpc.ncep.noaa.gov/index.htm The National Oceanic and Atmospheric Administration (NOAA) maintains the Climate Prediction Center. This site provides NOAA forecasts of weather-related extreme events, such as droughts, El Nino/La Nina events, and hurricanes.

Is the current or upcoming hurricane season in the Atlantic Ocean and Caribbean Sea predicted to be above or below average activity?

Is there an above or below average risk of hurricane landfall?

Have El Nino/La Nina advisories been issued? If so, what is the likely climatic impact on your location?

List other threats noted in the *U.S. Threats Assessments*.

Droughts

http://enso.unl.edu/monitor/monitor.html This site provides a summary of current drought conditions for the U.S. Based on the Drought Monitor Map, describe drought conditions at your location.

Lat. _____; Long. _____

Select *Current Conditions* for maps showing drought indices, precipitation, soil moisture, and snow pack conditions. For your location (or locations selected by your lab instructor), are the following measurements or indices above or below average, and to what extent:

Palmer Drought Severity Index

Crop Moisture Index

Percent of Normal Rainfall

Daily Streamflow

Snowpack (for mountain areas of the western U.S.)

LAB EXERCISE
10

Climate Classification

There are several climate classification systems used to describe similar climatic regions throughout the world. These classifications are based on different criteria, such as monthly or yearly precipitation and temperature, solar radiation, and length of the growing season. One of the most common climate classifications used by geographers and other earth scientists was developed by W. Koppen early this century and later revised by others.

Koppen's climate classification is shown in Figure 10.1. It is based on the global distribution of different types of vegetation. The type of vegetation in a region is determined primarily by climate, and boundaries for climatic regions and vegetation regions closely match. Koppen's climatic regions are designated by a series of letters. There are five basic climatic types. The general characteristics of each are listed below.

A - Tropical Humid Climates
B - Dry Climates
C - Warm Temperate or Subtropical Climates
D - Cool Temperate Climates
E - Polar Climates

Subtypes are designated by second and third letters which indicate differences in temperature and precipitation. Monthly and annual temperature and precipitation averages are required to use this classification. An overview of the classification is shown in Table 10.1.

Figure 10.1   Map of Koppen's Climate Classification.

Table 10.1 Koppen Climate Classification System

1st	2nd	3rd	Description and Criteria
A			**Tropical wet** (monthly temperature all greater than or equal to 18°C)
	f		Wet all year (every month receives 60mm or more of precipitation)
	m		Monsoon (short dry season — see Figure 10.3)
	w		Savanna (dry winter — see Figure 10.3)
B			**Dry** (potential evapo-transpiration exceeds precipitation — see Figure 10.2)
	W		Desert (see Figure 10.2)
	S		Steppe (see Figure 10.2)
		h	Tropical or hot (annual temperature is 18°C or greater)
		k	Mid-latitude or cool (annual temperature less than 18°C)
C			**Warm temperate** (average temperature of coldest month at least –3°C but less than 18°C; warmest month at least 10°C)
	w		Dry winter (driest winter month less than or equal to 1/10 of precipitation in wettest summer month)
	s		Dry summer (driest summer month less than or equal to 1/3 of precipitation in wettest winter month)
	f		Wet all year (all months receive at least 30mm precipitation)
		a	Hot summer (warmest month greater than or equal to 22°C)
		b	Warm long summer (at least 4 months have temperatures of 10°C or above but under 22°C)
		c	Short summer (1 to 3 months have temperatures of 10°C or above but under 22°C)
D			**Cool temperate** (coldest month less than –3°C; warmest month at least 10°C)
	w		same as C climate
	f		same as C climate
		a	same as C climate
		b	same as C climate
		c	same as C climate
		d	Severe winter (coldest month less than –38°C)
E			**Polar** (every month less than 10°C)
	T		Tundra with growing season (warmest month at least 0°C)
	F		Ice with no growing season (all months less than 0°C)

Procedure for Classifying Climates

To use Koppen's classification use the following procedure that tests for first letter classification in the following order: E, B, A, C, and then D. After establishing the first letter, determine the second letter classification, and then third letters for B, C, and D climates. Figure 10.2 is used to determine second letter B climates and to distinguish B climates from A, C, or D climates. Figure 10.3 is used to determine the second letter for A climates.

You must know if the climate station is in the northern or southern hemisphere. If you do not know the position of the climate station, the hemisphere can be determined by comparing monthly temperatures. If the coldest months are December, January, and February, it is in the northern hemisphere. If the coldest months are June, July, and August, it is in the southern hemisphere.

Note: T and **P** refer to average monthly values for temperature and precipitation.

1. If every T is less than 10°C, it is an **E** climate. If it is not an **E** climate, go to step 2.

   If every T is less than 0°C, it is an **EF** climate, if not it is an **ET** climate.

2. Use Figure 10.2 to determine if it is a **B** climate. If it is not a **B** climate go to step 3.

   a. Use Figure 10.2 to determine if it is a **BW** or **BS** climate.
   b. If annual temperature is less than 18°C, the third letter is k, either **BWk** or **BSk**. If annual temperature is greater than 18°C, the third letter is h, either **BWh** or **BSh**.

3. If every T is greater than or equal to 18°C, it is an **A** climate. If it is not an **A** climate go to step 4.

   a. If every P is greater than 60mm, it is **Af**.
   b. Use Figure 10.3 to determine is it is either **Am** or **Aw**.

4. If every T is greater than or equal to –3°C, it is a **C** climate. If it is not a **C** climate, go to step 5.

   Second letters

   a. If every P is greater than or equal to 30mm it is **Cf**.
   b. If the driest winter P is less than or equal to 1/10 of the wettest summer P it is **Cw**.
   c. If the driest summer P is less than or equal to 1/3 of the wettest winter P it is **Cs**. If not, it is **Cf**.

   Third letters

   d. If the warmest T is greater than or equal to 22°C, the third letter is **a**, either **Cfa**, **Cwa**, or **Csa**.
   e. If at least 4 T are greater than or equal to 10°C, the third letter is **b**, either **Cfb**, **Cwb**, or **Csb**.
   f. If fewer than 4 T are greater than or equal to 10°C, the third letter is c, either **Cfc** or **Cwc**. (**Csc** does not exist.)

5. It is a **D** climate.

Second letters

a. If every P is greater than or equal to 30mm it is **Df**.
b. If the driest winter P is less than or equal to 1/10 of the wettest summer P it is **Dw**, if not, it is **Df**.

Third letters

c. If the warmest T is greater than or equal to 22°C, the third letter is **a**, either **Dfa** or **Dwa**.
d. If at least 4 T are greater than or equal to 10°C, the third letter is **b**, either **Dfb** or **Dwb**.
e. If fewer than 4 T are greater than or equal to 10°C, the third letter is c, either **Dfc**, **Dwc**. (**Csc** does not exist.)
f. If the coldest month is less than −38°C, the third letter is **d**, either **Dfd** or **Dwd**.

"B" CLIMATE SUBTYPES

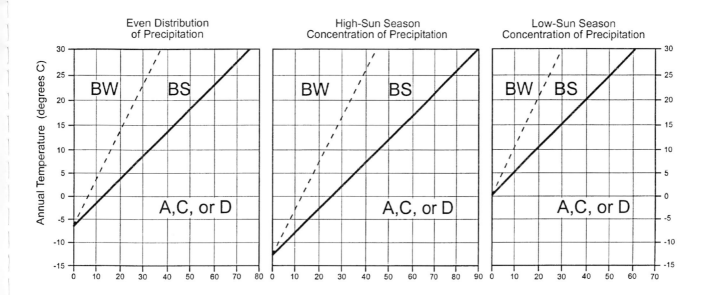

Annual Precipitation (cm)

Figure 10.2   Determine second letter B climates and distinguish B climates from A, C, or D climates.

Figure 10.2 is used to determine second letter B climates and to distinguish B climates from A, C, or D climates. To use these diagrams, follow these procedures:

1. Determine if the climate station is in the northern or southern hemisphere by comparing monthly temperatures.
2. Determine whether most of the precipitation occurs in the summer (high sun season), winter, (low sun season), or is evenly distributed throughout the year. For the purpose of this classification, the year is divided into two six month seasons, April–September and October–March.
3. If 70 percent or more of the annual total precipitation is in one of these six month periods, that season is said to have a precipitation concentration.
4. Use the appropriate one of these three diagrams and plot annual total precipitation and annual temperature values. If a point is plotted on the right side of the graph, it is not a dry climate, and is either an A, C, or D climate. If the point is plotted on the left side of the graph, it is a B climate.
5. If the climate station is classified as a B climate, use the graph to determine if it is a BS (desert steppe) or BW (true desert) which is drier.

"A" CLIMATE SUBTYPES

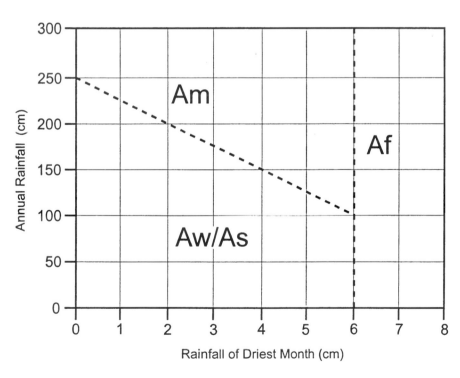

Figure 10.3　Determine the second letter for A climates.

104

Determine the Koppen classification for the following climate stations:

Note: Temperature is degrees C, precipitation is in mm. **Yr** values for temperature is an average of all the months. **Yr** values for precipitation are yearly totals.

Atlanta, Georgia, USA
Location: 34°N; 84°W. Elevation: 308m

	Jan	Feb	Mar	Apr	May	Jun	Jul	Aug	Sep	Oct	Nov	Dec	Yr
T	6	7	11	16	21	25	26	25	23	17	11	7	16
P	143	136	148	131	114	105	144	110	84	92	103	121	1437

Climate classification: _____

Cape Town, South Africa
Location: 34°S; 19°E. Elevation 12m

	Jan	Feb	Mar	Apr	May	Jun	Jul	Aug	Sep	Oct	Nov	Dec	Yr
T	21	21	20	17	15	13	12	13	14	16	18	20	17
P	16	15	21	50	92	105	91	82	54	40	24	19	612

Climate classification: _____

Chicago, Illinois, USA
Location: 42°N; 88°W. Elevation: 205m

	Jan	Feb	Mar	Apr	May	Jun	Jul	Aug	Sep	Oct	Nov	Dec	Yr
T	-4	-2	4	10	16	21	24	24	20	14	6	-1	11.0
P	48	40	73	91	99	93	91	76	65	69	61	61	905

Climate classification: _____

Beijing, China
Location: 40°N; 116°E. Elevation: 54m

	Jan	Feb	Mar	Apr	May	Jun	Jul	Aug	Sep	Oct	Nov	Dec	Yr
T	-4	-1	4	13	20	24	26	25	20	13	4	-3	11
P	3	4	8	18	33	78	224	170	58	18	9	2	635

Climate classification: _____

Phoenix, Arizona, USA
Location: 33°N; 112°W. Elevation: 337m

	Jan	Feb	Mar	Apr	May	Jun	Jul	Aug	Sep	Oct	Nov	Dec	Yr
T	11	13	16	20	25	30	33	32	28	22	15	11	21
P	16	20	19	6	3	1	16	29	22	14	14	27	189

Climate classification: _____

Ho Chi Min City, Vietnam
Location: 11°N; 107°E. Elevation: 19m

	Jan	Feb	Mar	Apr	May	Jun	Jul	Aug	Sep	Oct	Nov	Dec	Yr
T	27	27	28	29	29	28	27	27	27	27	26	26	27
P	14	4	9	51	213	309	295	271	342	261	119	47	1903

Climate classification: _____

Fairbanks, Alaska, USA
Location: 65°N; 148°W. Elevation: 189m

	Jan	Feb	Mar	Apr	May	Jun	Jul	Aug	Sep	Oct	Nov	Dec	Yr
T	-21	-18	-10	-1	9	15	16	13	7	-4	-15	-20	-3
P	15	13	10	7	17	45	56	53	32	22	19	21	310

Climate classification: _____

Paris, France
Location: 49°N; 2°E. Elevation: 66m

	Jan	Feb	Mar	Apr	May	Jun	Jul	Aug	Sep	Oct	Nov	Dec	Yr
T	3	4	6	10	14	17	19	18	15	11	6	3	11
P	46	39	41	44	56	57	57	55	53	57	54	49	607

Climate classification: _____

Tropical Climates (A)

Tropical climates are warm and humid and occur mostly within 20° of the equator. These regions do not have a winter season and the annual temperature range is small. Annual precipitation in most tropical regions is high. The year-round growing season and high precipitation support forest or mixed forest-grassland vegetation.

Most rainfall is caused by convection related to the Intertropical Convergence Zone (ITCZ). Migration of the ITCZ with changing declination of the sun causes seasonal variability in precipitation. Precipitation is also caused by (1) convection related to low pressure troughs in the region of the trade winds, referred to as *easterly waves*, and (2) tropical cyclones.

There are three subtypes of Tropical climates. First, Tropical Wet (Af) climates have year-round precipitation related to the ITCZ, and therefore, no dry season. Tropical Savannas (Aw) are under alternating influence of the ITCZ (wet season) and subtropical high pressure (dry season). Tropical Monsoons (Am) occur along tropical coasts backed by highlands. Like the Aw climates, these regions have alternating wet summer and dry winter seasons. However, Tropical Monsoons have a comparatively long and heavy rainy wet summer and a short dry winter.

Bangui, Central African Republic
Location: 4°N; 19°E. Elevation: 365m

	Jan	Feb	Mar	Apr	May	Jun	Jul	Aug	Sep	Oct	Nov	Dec	Yr
T	25	26	27	27	26	25	24	24	25	25	25	25	25
P	19	38	106	132	163	142	180	224	190	201	92	28	1507

Climate classification: _____

Kinshasa, Democratic Republic of the Congo
Location: 4°S; 15°E. Elevation: 309m

	Jan	Feb	Mar	Apr	May	Jun	Jul	Aug	Sep	Oct	Nov	Dec	Yr
T	25	25	25	25	25	22	22	22	25	25	25	25	25
P	138	148	184	219	148	5	3	4	40	133	235	156	1406

Climate classification: _____

Mergui, Myanmar
Location: 12°N; 99°E. Elevation: 2m

	Jan	Feb	Mar	Apr	May	Jun	Jul	Aug	Sep	Oct	Nov	Dec	Yr
T	26	27	28	28	28	27	26	26	26	27	27	26	27
P	22	50	77	131	438	772	808	768	619	312	88	19	4092

Climate classification: _____

Manaus, Brazil
Location: 3°S; 60°W. Elevation 65m

	Jan	Feb	Mar	Apr	May	Jun	Jul	Aug	Sep	Oct	Nov	Dec	Yr
T	26	26	26	26	26	26	26	27	27	28	27	27	27
P	264	262	298	282	204	103	67	46	63	111	161	220	2088

Climate classification: _____

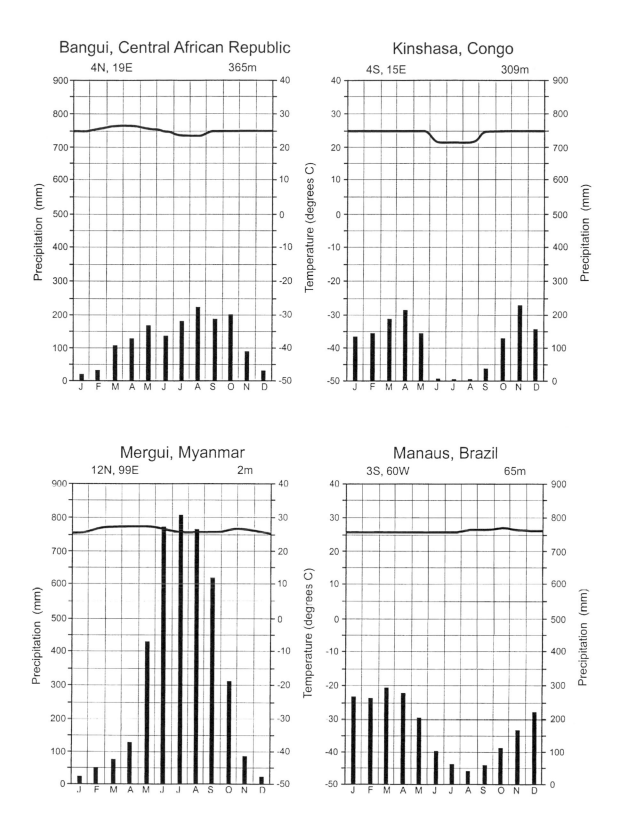

Figure 10.4   Tropical climates.

1. What is the annual temperature range for each location?

   Bangui, Central Africa Republic: _____

   Kinshasa, Democratic Republic of the Congo: _____

   Mergui, Mayanmar: _____

   Manaus, Brazil: _____

2. The temperature range for the tropics is small compared to higher latitudes. Why?

3. Do these locations have higher precipitation in the summer or winter? Explain this climatic pattern.

Dry Climates (B)

Desert and Steppe climate regions are defined by lack of water. They cover about 30 percent of the world's land area. These regions occur from low to high latitudes, and temperatures range from cold to hot. No permanent streams originate in these regions. Although permanent streams with headwaters in more humid regions will flow through these dry regions. True deserts (BW) support sparse or no vegetation except along streams. Steppe regions (BS) have somewhat more precipitation, although still too dry to support trees. Dry climates are the result of (1) rainshadow effect on the lee side of mountain ranges, (2) dominance of subtropical high pressure, (3) remoteness from moisture sources in the interior of large northern hemisphere continents, and (4) cold ocean current off the west coast of continents.

The daily temperature range in dry climates is often very high. Typically, there is little cloud cover and humidity is low. Also, there is little evaporative cooling of the surface. Daytime surface temperature can be very high. In contrast, temperatures can fall rapidly after sunset because of the rapid loss of longwave radiation.

Cairo, Egypt
Location: 30°N; 31°E. Elevation: 32m

	Jan	Feb	Mar	Apr	May	Jun	Jul	Aug	Sep	Oct	Nov	Dec	Yr
T	13.8	15.2	17.4	21.4	24.7	27.3	27.9	27.9	26.3	23.7	19.1	15.1	21.7
P	5	4	4	2	1	0	0	0	0	1	3	6	26

Climate classification: _____

Alice Springs, Australia
Location: 24°S; 134°E. Elevation: 537m

	Jan	Feb	Mar	Apr	May	Jun	Jul	Aug	Sep	Oct	Nov	Dec	Yr
T	29	28	25	20	16	12	12	14	18	23	26	28	21
P	41	42	35	17	17	17	12	10	9	20	25	37	281

Climate classification: _____

Iquiqeu, Chile
Location: 21°S; 70°W. Elevation: 10m

	Jan	Feb	Mar	Apr	May	Jun	Jul	Aug	Sep	Oct	Nov	Dec	Yr
T	21	21	20	18	17	16	16	15	16	17	18	20	18
P	0	0	0	0	0	1	1	0	0	0	0	0	2

Climate classification: _____

Death Valley, California, USA
Location: 36°N; 117°W. Elevation –59m

	Jan	Feb	Mar	Apr	May	Jun	Jul	Aug	Sep	Oct	Nov	Dec	Yr
T	11	15	19	24	30	35	38	37	32	25	17	10	24
P	8	12	9	4	2	1	3	3	4	3	6	4	60

Climate classification: _____

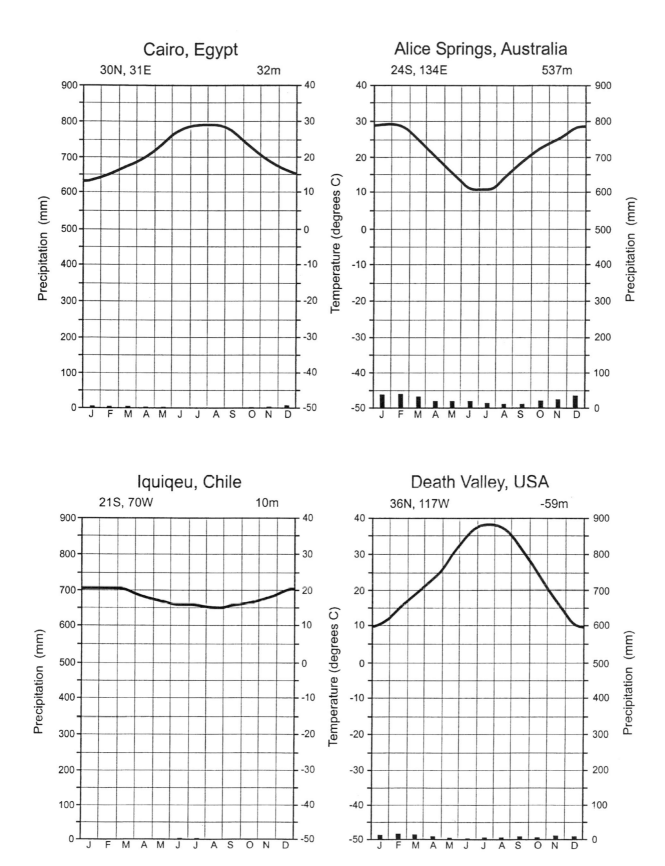

Figure 10.5   Dry climates.

1. What is the annual temperature range for each location?

   Cairo, Egypt: _____

   Alice Springs, Australia: _____

   Iquiqeu, Chile: _____

   Death Valley, California, USA: _____

2. What is the relationship between latitude and annual temperature range?

3. What are the dominant climatic controls causing dry conditions in each of these locations?

   Cairo, Egypt:

   Alice Springs, Australia:

   Iquiqeu, Chile: **Note:** This location is on the west coast of South America.

   Death Valley, California, USA:

Mid-Latitude Climates (C and D)

Mid-latitude climates are regions of conflicting air masses with alternating incursions of polar and tropical air. They are characterized by distinct seasonal changes in temperature and precipitation. Winter temperatures range from bitter cold in the interior of Asia and North America to mild along the coasts at lower latitudes. Summer temperatures range from hot to mild.

The dominant precipitation mechanisms are cyclonic storms and fronts. These are most common during the winter and spring. Convection is a common cause of precipitation during the summer, especially on the southeastern parts of continents because of high summer-time humidity in these regions.

Warm Temperate climates (C) occur mostly on the western and southeastern portions of continents. Cold continental climates (D) cover much of the high and mid-latitudes of the large continents of the northern hemisphere. Cold continental climates do not occur in the southern hemisphere because of the absence of large land masses in the mid-latitudes.

London, England
Location: 52°N; 0°W. Elevation: 24m

	Jan	Feb	Mar	Apr	May	Jun	Jul	Aug	Sep	Oct	Nov	Dec	Yr
T	5	5	7	9	13	16	18	18	15	12	8	6	11
P	62	36	50	43	45	46	46	44	43	73	45	59	588

Climate classification: _____

Athens, Greece
Location: 38°N; 24°E. Elevation: 28m

	Jan	Feb	Mar	Apr	May	Jun	Jul	Aug	Sep	Oct	Nov	Dec	Yr
T	10	11	12	16	21	25	28	28	24	19	15	12	19
P	48	41	41	23	18	7	5	8	10	53	55	62	371

Climate classification: _____

Fuzhou, China
Location: 26°N; 119°E. Elevation: 152m

	Jan	Feb	Mar	Apr	May	Jun	Jul	Aug	Sep	Oct	Nov	Dec	Yr
T	11	11	14	18	22	26	29	29	26	22	18	14	20
P	47	87	120	139	179	210	138	172	165	47	40	38	1389

Climate classification: _____

Irkutsk, Russia
Location: 64°N; 100°E. Elevation: 650m

	Jan	Feb	Mar	Apr	May	Jun	Jul	Aug	Sep	Oct	Nov	Dec	Yr
T	-36	-32	-19	-7	3	12	16	12	5	-7	-26	-33	-9
P	13	10	9	13	25	51	62	54	36	25	20	16	335

Climate classification: _____

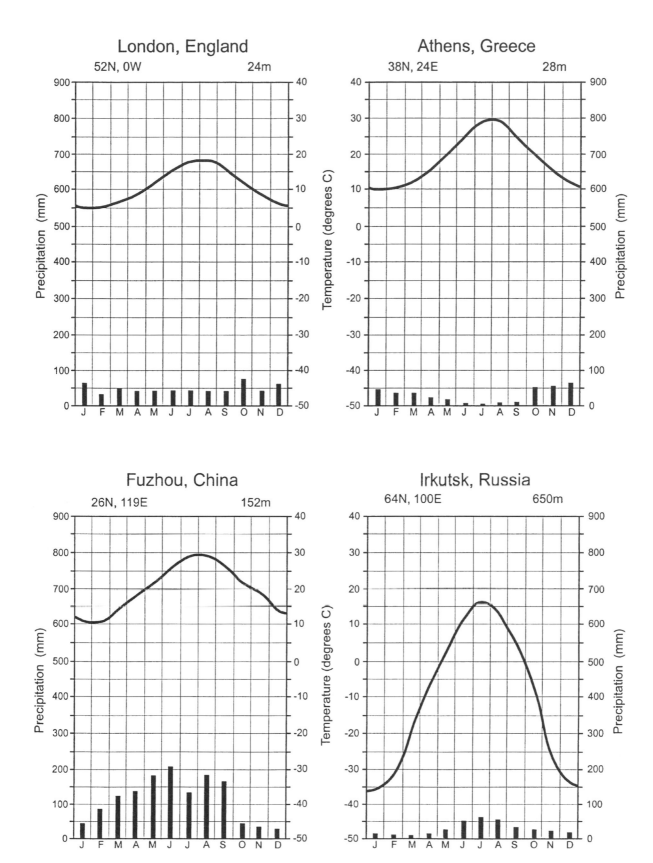

Figure 10.6   Mid-latitude climates.

1. What is the annual temperature range for each location?

    London, England: _____

    Athens, Greece: _____

    Fuzhou, China: _____

    Irkutsk, Russia: _____

2. London, England is located 52°N, a higher latitude than all areas in the continental U.S. However, London has mild winter temperatures and a low annual temperature range. Explain why this location has moderate temperatures.

3. Irkutsk, Russia is located in central Asia. Explain why this location has an extremely high annual temperature range.

4. What climatic control explains the distinct wet winter and dry summer occurring in Athens, Greece?

5. Mid-latitude cyclones and fronts occur more frequently during the winter and these are the dominant precipitation mechanisms for most of the mid-latitudes. Why does Fuzhou, China have a wet summer season?

Climate Classification: Internet Resources

Climate Data

http://www.worldclimate.com has climate data from thousands of sites throughout the world. These include monthly and yearly averages for temperature, precipitation, and, pressure.

Note the following climate information at your location:

Place name: _____

Lat. _____; Long. _____

      Average Annual Temperature: _____°C

      Annual Temperature Range (warmest month-coldest month): _____°C

      Coldest Month: _____ , Temperature: _____°C

      Warmest Month: _____ , Temperature: _____°C

      Average Annual Rainfall: _____ mm

      Wettest Month: _____ , Rainfall: _____ mm

      Driest Month: _____ , Rainfall: _____ mm

Climate Classification: _____

Use the **navigation** tool at the bottom of the page to find other climate recording stations. Select the compass direction (**North**, **North-East**, **East**, etc.) you wish to move (or selected by the lab instructor).

1.   Place name: _____

    Lat. _____ ; Long. _____

      Average Annual Temperature: _____°C

      Annual Temperature Range (warmest month-coldest month): _____°C

      Coldest Month: _____ , Temperature: _____°C

      Warmest Month: _____ , Temperature: _____°C

      Average Annual Rainfall: _____ mm

Wettest Month: _____ , Rainfall: _____ mm

Driest Month: _____ , Rainfall: _____ mm

Climate Classification: _____

2. Place name: _____

   Lat. _____ ; Long. _____

   Average Annual Temperature: _____°C

   Annual Temperature Range (warmest month-coldest month): _____°C

   Coldest Month: _____ , Temperature: _____°C

   Warmest Month: _____ , Temperature: _____°C

   Average Annual Rainfall: _____ mm

   Wettest Month: _____ , Rainfall: _____ mm

   Driest Month: _____ , Rainfall: _____ mm

Climate Classification: _____

What is the proximity of these climate recording stations to your location (or position first noted above)?

How has the change in latitude or distance from the coast relative to your recording station (or the one first listed above) affected the climate?

Name _____ Section _____

LAB EXERCISE
11

Topographic Map Interpretation

Topographic maps are among the most useful tools for earth scientists. They show physical features of the earth's surface (such as elevation, streams, glaciers, vegetation), man-made features (buildings, roads, mines), and geographical grid systems that are used to determine location. The purpose of this exercise is for you to become familiar with the features of these maps, including map symbols, scale, and contour lines. It is important that you are able to interpret topographic maps since they will be used extensively with the later exercises in this lab manual.

Topographic maps use a standard set of symbols to identify both natural and human features of the landscape. A legend of these symbols is placed in the back of this text. Since you will be using these maps extensively in the following exercises you need to be familiar with map symbols.

Sections of ten topographic maps are included in this manual to illustrate a variety of landscapes. The map scale and contour intervals are noted on the back of each. These maps are only portions of the printed originals and do not include information included in the margins (such as publication date, magnetic declination, grid coordinates, and names of adjoining maps).

A complete quadrangle should be available for students at the beginning of this exercise. The map sections that are included will be sufficient for the remainder of exercises in this manual. But the lab instructor may prefer to use maps that illustrate features in your area.

A topographic map will be provided by your lab instructor.

1. The map _____
                         (name of quadrangle)

2. Geographic grid coordinates at the lower right corner:

    _____

3. Date of map: _____

4. Scale (Relative Fraction): _____

5. Contour interval: _____ feet

6. What quadrangle joins this on the south? _____

7. What information is represented by these colors?

Blue: _____

Green: _____

Red: _____

Map Scale

One of the most important features of a map is its scale. Map scale is the ratio between the distance on the map and the corresponding distance on the ground. The ability to work with map scale is necessary for proper interpretation.

The most commonly used scale is **Representative Fraction** (R.F.) which is usually shown as a ratio:

distance on map: distance on ground

An R.F. of 1:12,000 means that one inch on the map equals 12,000 inches on the ground. Of course, different units of measurement (such as cm or feet) can be used, but you must use the same units on both sides of the ratio.

Conversion of Units

You can rapidly convert from one scale unit to another. Suppose you want to convert 12,000 inches to miles. Setup the problem so that you multiply the number of inches (12,000) by 1 mile divided by its equivalent in inches. There are 63,360 inches per mile. So:

$$12{,}000 \text{ inches} \quad * \quad \frac{1 \text{ mile}}{63{,}360 \text{ inches}}$$

Inches cancel on the top and bottom so that only the mile unit remains. Then multiply:

$$12{,}000 \text{ inches} \quad * \quad \frac{1 \text{ mile}}{63{,}360 \text{ inches}} \quad = \quad \frac{12{,}000 \text{ mile}}{63{,}360} \quad = \quad 0.189 \text{ miles}$$

Examples

1. Convert 800 inches to feet:

$$800 \text{ inches} \quad * \quad \frac{1 \text{ foot}}{12 \text{ inches}} \quad = \quad \frac{800 \text{ feet}}{12} \quad = \quad 66.6 \text{ feet}$$

2. Convert 1600 feet to miles:

$$1600 \text{ feet} \quad * \quad \frac{1 \text{ mile}}{5280 \text{ feet}} \quad = \quad \frac{1600 \text{ miles}}{5280} \quad = \quad 0.30 \text{ miles}$$

You can also use this method to easily convert between English and metric units:

3. Convert 1334 centimeters to inches:

$$1334 \text{ cm} \quad * \quad \frac{1 \text{ inch}}{2.54 \text{ cm}} \quad = \quad \frac{1334 \text{ inches}}{2.54} \quad = \quad 525.2 \text{ inches}$$

4. Convert 22.3 miles to kilometers:

$$22.3 \text{ miles} \quad * \quad \frac{1 \text{ km}}{0.621 \text{ miles}} \quad = \quad \frac{22.3 \text{ km}}{0.621} \quad = \quad 35.9 \text{ km}$$

English measurements can be confusing because the corresponding units do not have consistent conversion ratios (12 inches = 1 foot, 5280 feet = 1 mile). This is one of the major reasons the U.S. is switching to the metric system, which is used by most of the world's industrialized countries. Metric units can be divided into even numbers of 10, 100, or 1000.

1 centimeter (cm)	=	10 millimeters (mm)
1 meter (m)	=	100 centimeters
1 kilometer (km)	=	1000 meters

Example

With an R.F. of 1:12,000, one centimeter on the map = 12,000 centimeters on the ground. The distance on the ground (in centimeters) can be converted to meters or kilometers:

$$12,000 \text{ cm} \quad * \quad \frac{1 \text{ m}}{100 \text{ cm}} \quad = \quad \frac{12,000 \text{ m}}{100} \quad = \quad 120 \text{ m}$$

$$12,000 \text{ cm} \quad * \quad \frac{1 \text{ km}}{100,000 \text{ cm}} \quad = \quad \frac{12,000 \text{ km}}{100,000} \quad = \quad 0.12 \text{ km}$$

The **Seagrove** map shows an area of central North Carolina. The map has a Relative Fraction (R.F.) of 1:24,000. What is the distance on the ground between the Fairview Church, southeast of the town of Seagrove and the Fulp Church, north of Seagrove.

1. First, measure the distance on the map between these two points.

    _____ (cm or inches)

2. Multiply this value by the denominator of the RF to determine the distance on the ground.

    _____

3. Convert this distance to either feet or meters, depending on if you are using English or Metric units.

    _____ (feet or meters)

    Enter this value below and convert to the other units shown.

_____ feet

_____ miles

_____ meters

_____ kilometers

4. What is the distance across the Seagrove map? _____

Refer to the maps located in different chapters for the following questions.

5. **Heidelberg, Kentucky** (Lab Exercise 17)

    R.F.:_____

    Width of the Kentucky River at the mouth of Cave Branch:

    _____

6. **Cape Solitude, Arizona** (Lab Exercise 20)

    R.F.:_____

    Distance from Sixtymile Rapids to the mouth of the Little Colorado River:

    _____

7. **Fishhook Lake, Mississippi** (Lab Exercise 20)

    R.F.:_____

    Shortest distance between Black Lake and the Tallahatchie River:

    _____

8. **Stovepipe Wells, California** (Lab Exercise 20)

    R.F.:_____

    State the verbal scale of the map:

    a. inches to miles

        Scale: _____ inch = _____ mile

    b. centimeters to kilometers

        Scale: _____ cm = _____ km

Seagrove, North Carolina

Scale   1:24,000

Contour Interval 10 Feet

Calculating Area

The R.F. can be used to determine area. For example, if the R.F. = 1:24,000, one inch on the map equals 24,000 inches on the ground, or 2,000 feet. Therefore, 1 square inch on the map or photograph (1x1), the area on the ground would be calculated as:

$$2000 \times 2000 = 4,000,000 \text{ feet}^2$$

1. Assume a map has a scale of 1: _____ . If the area on the topographic map is ____ x ____ (in. or cm), determine the corresponding area on the ground.

    _____ feet^2

    _____ mile^2

    _____ meters^2

    _____ km^2

2. Assume a map has a scale of 1: _____ . If the area on the topographic map is ____ x ____ (in. or cm), determine the corresponding area on the ground.

    _____ feet^2

    _____ mile^2

    _____ meters^2

    _____ km^2

Circular areas can also be easily calculated. Refer to Figure 11.1 shown below.

Figure 11.1   Circular area.

3. To determine the area of a circle in this problem, use the following R.F.: _____.
    The side of the square is _____ inches, which is also the diameter of the circle.

    The diameter of the circle on the ground = _____ feet

    The area of the circle = 3.14 * radius^2

    The area of the circle = _____ feet^2

4. Refer to the **Mt. Capulin** aerial photograph in Lab Exercise 19. The R.F. is 1:20,000.

What is area of the cone? _____

What is the area of the crater? _____

Topographic Contours

A contour line is a line on a map that connects points having equal elevation, or distance above (or below) sea level. These are among the most useful features of maps because they show how the landscape varies and can be used to identify different types of landforms.

The elevational distance between two contour lines is the **contour interval**. The contour interval used depends primarily on the map scale and the relief of the topography. Intervals of 5, 10, or 20 feet are common.

Contour Line Characteristics

1. A contour line represents only one elevation.
2. Contour lines do not divide or split.
3. Contour lines do not end. They will reconnect, but may run off maps that represent only a small portion of the landscape.
4. Contours are closer together on steep slopes than gentle slopes, but they never cross one another. On vertical slopes contour lines will overlap.
5. Contour lines bend upstream into an inverted "V" when crossing streams, and bend downslope when crossing ridges.
6. The contour interval is constant unless otherwise noted.
7. Hatchured contours indicate a depression. The hatchured lines have small tick marks on the downhill side.

Figure 11.2 shows contour lines and the corresponding landscape profile.

The south and western part of the **Platte City** map includes a portion of the Missouri River and Platte River floodplain. The wide spacing of contour lines in these areas indicate that these surfaces are nearly flat. In contrast, the upland areas adjacent to the floodplain have a more rugged terrain. The closely spaced contour lines on this section of the map indicate steeper slopes.

Answer the following using the **Platte City** map:

1. Contour interval: _____

2. Elevation of Green Cemetery is _____ ft

3. Elevation of _____ is _____ ft

4. Elevation of _____ is _____ ft

5. What is the location of the highest point on the map? (name or coordinates)

What is the elevation of that point? _____

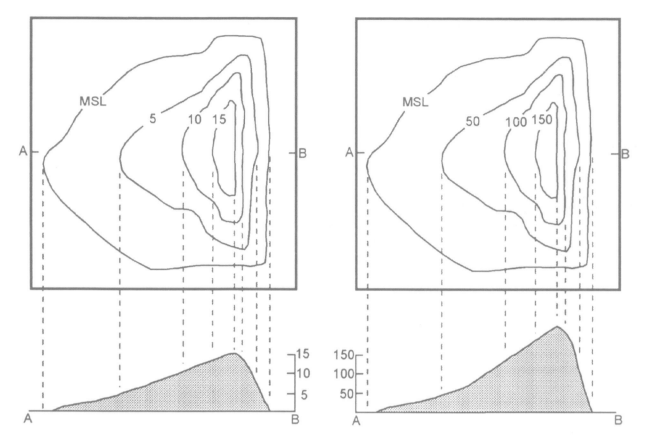

Figure 11.2   Landscape profile and corresponding contours.

6. What is the location of the lowest point on the map? (name or coordinates)

    What is the elevation of that point? _____

7. **Relief** is the difference between the highest and lowest points of a specific area. What is the relief for the area shown on this map?

    _____

8. What area of the map has the steepest slopes? (name or coordinates)

9. In which direction does the Platte River flow? _____

10. In what direction does the bluff on which the town of Farley occurs face?

    _____

11. Why do you think the town of Farley was built on the bluff rather than in the floodplain?

Drawing Contour Lines

On Figure 11.4 is a map with elevational points. The 170 foot contour has already been drawn. Draw the remaining contour lines at intervals of 10 feet. Be sure that all points with a lower elevation than the contour line you are drawing are on the downhill side of the line. Use a pencil and draw lightly until you are sure the contour positions are correct, then darken the lines. Be sure the inverted "V"s coincide with the stream channels.

Slope Gradient

The slope of the land surface can easily be determined on contour maps. The gradient, or slope angle, is often expressed as a ratio:

$$\frac{\text{rise}}{\text{run}}$$

where:

    rise = the elevational difference between 2 points
    run   = the horizontal distance between them.

    For example, on the cross-section of a hillslope shown in Figure 11.3, point A is upslope of point B.

Figure 11.3   Cross-section of a hillslope.

Difference in elevation between points A and B: 30 feet
horizontal distance between points A and B: 100 feet

The slope can be expressed in one of the following ways:

a.   Percent grade: 30%

      $\dfrac{30}{100}$ =   .30   =   30%

b.   Feet per mile: 1584 feet per mile

      $\dfrac{30}{100}$ = $\dfrac{x}{5280}$

         100x   =   158,400
             x   =   1584 feet per mile

Platte City, Missouri–Kansas

Scale   1:24,000

Contour Interval 10 Feet

Lake Contouria
Elevation 160 feet

Figure 11.4

Refer to the **Stovepipe Wells** map (Lab Exercise 20).

R.F.: _____

1. Determine the slope gradient between the mouth of Mosaic Canyon and the waste treatment plant south of Stovepipe Wells.

    rise (vertical distance) = _____ feet

    run (horizontal distance) = _____ feet

    a. Percent grade: _____

    b. Feet per mile: _____

    Stream Gradient can be calculated in the same way as percent grade, and is also expressed in feet per mile (or meters per kilometer). For example, you select a stream or stream segment that is 10 miles long (**run**). By using the contours at the lower and upper ends of the stream the vertical difference is 140 feet (**rise**). Therefore:

    $$\frac{140 \text{ ft elevational difference}}{10 \text{ mile distance}} = 14 \text{ ft per mile}$$

2. Refer to the **Heidelberg**, Kentucky map (Lab Exercise 18). Cave Branch is a small stream south of Mayse School that flows through rugged terrain before entering the Kentucky River. Determine the stream gradient of Cave Branch, taking into account bends in the stream and meanders when measuring stream length.

    Elevation difference (rise): _____ feet

    Stream length (run): _____ feet

    a. Percent grade: _____

    b. Feet per mile: _____

Refer to the **Ripley South** map. A dam is located on Hyde Creek, just south of Ripley, TN.

1. Approximately what is the water surface elevation of the reservoir?

    _____

2. Assume the dam is enlarged and heightened so that the water level reaches a height of 360 feet.

    a. How far from the dam will water backup into the valley?

    _____

    b. Approximately how many square miles will the surface of the reservoir cover?

    _____

134

c. Will any buildings be destroyed if the water rises to this elevation (360 feet)?

_____

How many?   _____

d. What is the length of the roadway that will be flooded?

_____ feet

Ripley South, Tennessee

Scale   1:24,000

Contour Interval 10 Feet

Topographic Map Interpretation: Internet Resources

Topographic Maps

http://www.topozone.com On this site you can view topographic maps for the entire United States. You can search for maps using place names and view different map scales.

http://www.usgs.gov/ The U.S. Geological Survey provides a site to search and order topographic maps that cover the entire lower 48 states. Click ***Topographic Maps*** to find USGS maps and order forms.

Trigonometry Applications in Geographical Field Work

Direct measurements of distance, area, or change in elevation, during fieldwork are often difficult or impractical. Trigonometry provides simple techniques to make indirect, but accurate calculations. In this exercise you will use (1) solutions for Right Triangles to determine both distance and height of objects and (2) solutions for Oblique Triangles to determine area.

Solutions for Right Triangles

Solutions for Right Triangles can be used to easily calculate heights of objects or changes in elevation. For example, you can use this technique to determine the height of trees, elevation of a cloud base, or elevation of a distant point relative to the elevation of your position. Refer to the Right Triangle shown below in Figure 12.1, where A, B, and C represent angles and a, b, and c are distances.

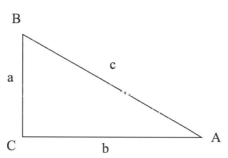

Figure 12.1   Right triangle.

Table 12.1 Solution for Right Triangles

Given	Determine	Using the Following Solutions
A, a	b	$b = a * \cot A$
	c	$c = a/\sin A$
A, b	a	$a = b * \tan A$
	c	$c = b/\cos A$
A, c	a	$a = c * \sin A$
	b	$b = c * \cos A$

Note: Trigonometry tables are given in Appendix B.

141

1. Determine height of an object or change in elevation in Figure 12.2.

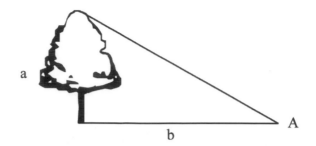

Figure 12.2   Determining height or change in elevation.

Assume you want to determine the height of a tree. Direct measurement usually is not possible. But you can use the solution for Right Triangles shown in Table 12.1. In this case, a is the height of the tree. You can easily measure your distance from the tree (b) and the angle from the ground to the treetop (A). Since you know A and b, use the solution shown above to determine tree height. In this case:

$$\text{Height of the tree } (a) = b * \text{tangent of A}$$

Distance to the tree (b)	Angle° (A)	Tangent of A	Height (a)
15m	17	.3057	

You can use the same solution to determine the height of a distant point (building, hill or cloud base). If you know the horizontal distance to that point (b) and measure the angle (A), use the same solution shown above to determine the change in elevation, again referring to Figure 12.1.

Horizontal distance (b)	Angle° (A)	Tangent of A	Elevation
1km	11		

2. Determine distance using right triangles.

Using right triangles on maps or the ground, you can determine the distance between points. Assume you are standing on the bank of a river and want to know the distance to the opposite river bank as shown in Figure 12.3. You can use the same solution used to calculate height.

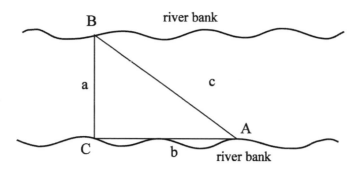

Figure 12.3   Using right triangles to determine distance.

In this case, a is the channel width. You can easily measure distance b on your side of the river and angle A. Use the following solution to determine distance across the river, as shown below:

$$\text{Channel Width } (a) = b * \text{tangent of A}$$

Distance along the bank (b)	Angle° (A)	Tangent of A	Channel width (a)
30m	47		

If you know the distance of any one side of the triangle, you can measure angle A and then use the solutions given in Table 12.1 to determine the distances of the other sides.

Given (a, b, or c)	Determine (a, b, or c)	Angle° (A)	Distance

Solution for the Area of Oblique Triangles

An oblique triangle is any triangle that is not a right triangle. It could be an acute triangle (all three angles less than 90°) or an obtuse triangle (one of the angles greater than 90°). Surface area on the ground or on maps can easily be determined by using oblique triangles. For example, you can calculate the area of any triangle on a map or on the ground by identifying the position of points A, B, and C, and measuring the distance between them. Refer to the oblique triangles shown below in Figure 12.4.

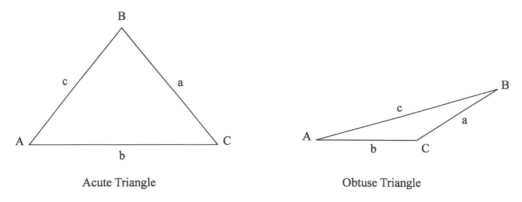

Acute Triangle   Obtuse Triangle

Figure 12.4   Oblique triangles.

Solution for the area of oblique triangles:

Given   Determine   Using the following solution

a,b,c   area   area = $\sqrt{s\,(s\text{-}a)\,(s\text{-}b)\,(s\text{-}c)}$ where: s = (a + b + c)/2

Assume you want to determine a triangular area on the ground or corresponding area on a map. Measure the distance of each side of the triangle and then calculate the area using the above solution. **Note:** If the area is not triangular, you can use points on the perimeter that allow you to divide the area in contiguous triangles.

	Distances		s	Area
a __15m__ b __12m__ c __11m__			_____	_____
a_____ b_____ c_____			_____	_____
a_____ b_____ c_____			_____	_____

Field Exercise

The lab instructor will indicate positions or objects for you to determine height (or elevation change), distance, or area using solutions for right and oblique triangles. Maps can be used if field observations are not possible.

Plot positions given by the lab instructor and triangles used to find solutions on the graph paper included at the end of this exercise. Draw the map with a pencil, not a pen, in case corrections are needed. The quality of your maps depend on accuracy! Use an engineers scale or ruler for all line work, and a protractor to determine angles.

1. Calculate distance using a right triangle solution. First determine distance (*b*) and then Angle (A).

	Your position (point of observation)		Determine distance to following positions or objects	
1. _____			_____	
2. _____			_____	
3. _____			_____	

	Distance (b)	Angle° (A)	Tangent of A	Distance (a)
1. _____		_____	_____	_____
2. _____		_____	_____	_____
3. _____		_____	_____	_____

2. Determine height (or elevation change) using right triangle solution. First, determine the horizontal distance to the position (b) and then angle A.

	Your position (point of observation)		Determine distance to following positions or objects	
1. _____			_____	
2. _____			_____	
3. _____			_____	

	Horizontal Distance (b)	Angle° (A)	Tangent of A	Distance (a)
1. _____		_____	_____	_____
2. _____		_____	_____	_____
3. _____		_____	_____	_____

3. Determine area using oblique triangle solution. First, you must identify and locate points on the perimeter of the designated area and create a triangle.

If the area is not triangular, use points on the perimeter that allow you to divide the area in contiguous triangles.

Determine the distance of each side of the triangle(s). Then find the area for each triangle and, if there is more than one triangle, add the values to determine the entire area.

Problem 1

Site description and boundary positions:

	Distances		s	Area
_a_____	_b_____	_c_____	_____	_____
_a_____	_b_____	_c_____	_____	_____
_a_____	_b_____	_c_____	_____	_____
_a_____	_b_____	_c_____	_____	_____
_a_____	_b_____	_c_____	_____	_____

Total Area _____
(if combining the area of more than one triangle)

Problem 2

Site description and boundary positions:

	Distances		s	Area
_a_____	_b_____	_c_____	_____	_____
_a_____	_b_____	_c_____	_____	_____
_a_____	_b_____	_c_____	_____	_____
_a_____	_b_____	_c_____	_____	_____
_a_____	_b_____	_c_____	_____	_____

Total Area _____
(if combining the area of more than one triangle)

147

Contour Mapping

Equipment

    3 brunton compasses with tripods
    3 stadia rods
    1 distance tape

Geographical field work occasionally requires the development of contour maps that show greater detail than the USGS topographic maps such as those used in Exercise 11. In this exercise you will collect field data from a small area designated by the lab instructor and use this information to create a contour map.

   A sample data set is included at the end of this exercise. The lab instructor may use this data if an appropriate field site is not available, or to demonstrate how contour maps are constructed before you begin data collection.

Data Collection

1. The class will be divided into three groups, each working at a different compass.

    Establish 2 points (Station A and Station B) _____ feet apart along a designated line (baseline). Three compasses will be used. At Station A, 2 compasses will be set up: one for measuring inclination to determine the change in elevation or altitude relative to your position (group 1), and the second for azimuth or compass direction (group 2). At Station B the third compass will be used to determine azimuth (group 3).

    The study site will be within a rectangular area (designated by the lab instructor) adjacent to the baseline as shown in Figure 13.1.

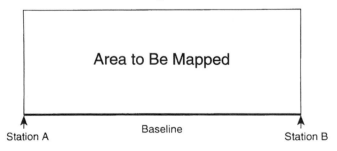

Figure 13.1   Study site.

151

2. One person from each group will take a stadia rod to a point within the study area. All three compasses will take readings based on this position. Azimuths will be determined from both Station A and Station B and entered on the following Table 13.1. The inclination (in degrees) will be determined from the third compass (at Station A) and also entered on Table 13.1 Everyone in each group will take a turn placing the stadia rod.

Table 13.1 Data Entry

Station A is _____ feet elevation.

Names (observation points)	Station A (azimuth)	Station B (azimuth)	Inclination (Station A) (degrees)
_____	_____	_____	_____
_____	_____	_____	_____
_____	_____	_____	_____
_____	_____	_____	_____
_____	_____	_____	_____
_____	_____	_____	_____
_____	_____	_____	_____
_____	_____	_____	_____
_____	_____	_____	_____
_____	_____	_____	_____
_____	_____	_____	_____
_____	_____	_____	_____
_____	_____	_____	_____
_____	_____	_____	_____
_____	_____	_____	_____
_____	_____	_____	_____
_____	_____	_____	_____
_____	_____	_____	_____
_____	_____	_____	_____
_____	_____	_____	_____
_____	_____	_____	_____

Mapping

1. Create a map using either the data collected in the field (or you can use the data in Table 13.3 at the end of this exercise). Plot the map on the graph paper included at the end of this exercise.

    On the graph paper, all vertical lines will be oriented north-south and horizontal lines will be east-west. Draw the map with a pencil, not a pen, in case corrections are needed. The quality of your map depends on accuracy! Use an engineers scale or ruler for all line work, and a protractor to determine angles.

    The scale for this map is _____ . Draw the baseline (a line from Station A to Station B). Remember that this line on the ground was _____ feet.

2. For each observation point, center the protractor on Station A and draw lines along the azimuths for the data. Repeat this using Station B. As shown in Figure 13.2, the intersection of each pair of lines determines the position of the point on the ground where the stadia rod was located.

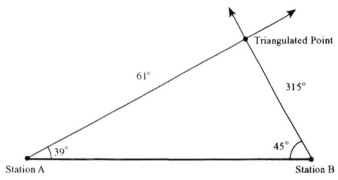

Figure 13.2   Mapping.

3. Measure the distance between all points and Station A and enter these values on the worksheet. These values should be calculated to the nearest 0.1 inch.

4. Use the map scale to determine the actual distance (X) on the ground. Enter the ground distances between all points and Station A on the worksheet.

5. Calculate the Y value (the difference in elevation from the compass that recorded inclination to a position 4 feet above the ground on the stadia rod). Determine the elevation of each point using the following:

Tangent of Angle a = Y/X
Therefore: Y = X * Tangent of Angle a

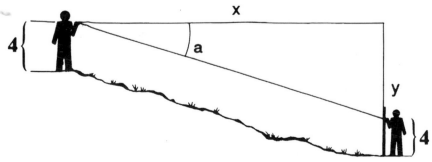

Figure 13.3   Mapping.

153

6. Construct a contour map by plotting the actual elevation for each survey point and replacing the point number with actual elevation. Draw contour lines using a contour interval of _____ ft., beginning with _____ ft. elevation.

The map must include a title, scale, date, direction orientation, and key (contour interval).

WORKSHEET

Point	Inclination	Measured Distance	Actual Distance(X)	Tangent	Y	Actual Elevation
1						
2						
3						
4						
5						
6						
7						
8						
9						
10						
11						
12						
13						
14						
15						
16						
17						
18						
19						
20						
21						
22						
23						
24						
25						

Sample Data Set

Table 13.2 is a sample data set that can be used to demonstrate how triangulation and inclination data can be used to construct a contour map. To use this data set, establish a north-south running baseline with Station A at the south-end of the baseline and Station B at the north-end. The study (mapped) area will be on the west side of the baseline.

Station A is _____ feet elevation

Refer to the mapping instructions on the preceding pages. A worksheet and graph paper for constructing the map are on the following pages.

Table 13.2 Sample Data Set

Observation Point	Azimuth A	Azimuth B	Inclination (Station A)
1	280	190	d9
2	300	200	d11
3	310	210	d11
4	320	230	d10
5	335	260	d12
6	0	180	d10
7	280	200	d11
8	280	210	d10
9	300	190	d10
10	300	210	d11
11	300	230	d8
12	310	190	d9
13	310	200	d12
14	310	230	d9
15	320	190	d10
16	320	200	d12
17	320	210	d11
18	335	190	d10
19	335	200	d11
20	335	210	d10
21	335	230	d11

WORKSHEET FOR SAMPLE DATA

Point	Inclination	Measured Distance	Actual Distance(X)	Tangent	Y	Actual Elevation
1						
2						
3						
4						
5						
6						
7						
8						
9						
10						
11						
12						
13						
14						
15						
16						
17						
18						
19						
20						
21						
22						
23						
24						
25						

157

159

Contour Mapping: Internet Resource

Elevation Maps of the United States

http://fermi.jhuapl.edu/states/states.html This site is a landform atlas of the United States. Click on any state to view either color or back and white elevation maps, or satellite images.

LAB EXERCISE
14

Soil Properties

Soil refers to the unconsolidated material occupying the earth's surface which provides a medium for plant growth. Physical and biological activity in the upper levels of soils results in characteristics that distinguish soils in one region from another. Five factors are considered essential in the development of soils: parent material, climate, vegetation, topography, and time. Parent materials are important to the physical and chemical makeup of young soils, but its influence is gradually reduced with increasing maturity.

Climate is the most important developmental factor in mature soils. Climate affects the rate of weathering, amount of chemical activity, the abundance of microorganisms, in addition to the vegetative cover and amount of organic material at the soil surface. Topography is also important in soil development. Slope and drainage can result in distinct soil characteristics even though parent material, climate, and vegetation are similar. Time is included among factors in soil development because of the direct relationship between the age of the development of soil profiles.

Soil Profiles

Soil profiles are sequences of horizontal soil layers called **soil horizons**. Horizons differ from those immediately above or below in physical, chemical, and biological properties. The major horizons are:

O: Organic horizons formed above mineral soil containing: (O_1) recognized forms of vegetation and (O_2) highly decomposed organic matter.

A: Mineral horizons consisting of: (A_1) a mixture of organic matter near the surface, (A_2) horizons that have been leached of clay, iron, and aluminum to lower horizons with a resulting concentration of sand and silt minerals, and (A_3) a lower zone similar to A_2, but transitional to lower horizons. The A horizon is a zone of **eluviation**, where soil material is moved downward to the lower horizons.

B: Horizons containing weathered material and an accumulation of iron and aluminum: (B_1) a zone transitional from A to B but containing material leached from above, (B_2) zone of maximum accumulation (**illuviation**), and (B_3) a lower zone similar to B_2 but transitional to C.

C: Zone of unconsolidated material, least weathering, and little organic material.

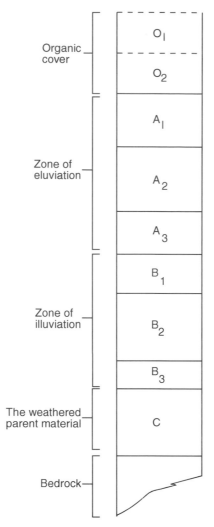

Figure 14.1   Idealized soil profile.

Soil profiles are highly variable depending on the 5 developmental factors listed above. There is a strong regional relationship between climate and vegetation, and both have a strong, direct effect on soil processes. Therefore, humid regions with forest vegetation typically have soils with similar properties, as do semi-arid grasslands, and deserts.

Soil Profile Descriptions

The lab instructor will determine an outdoor location to observe a soil profile. Record your visual observations on the data sheets at the end of this exercise. In addition, draw horizon boundaries, designations, and features on the graphs at the end of this exercise. On the drawings, include the depth (cm) of each horizon.

Afterwards, read the description of these soils from the County Soil Survey or information provided by the lab instructor and compare this information to your observations.

Physical Properties of Soils

The most important physical properties of soils are texture and structure. **Texture** is the proportion of different sized soil particles (sand, silt, and clay) and **structure** refers to the way in which the soil particles are grouped. Texture and structure determine the pore space in soils that affects rate of infiltration and water holding capacity, susceptibility to erosion, and ease of cultivation. Also, texture strongly affects chemical and biological processes. The size of soil particles are classified into three major groups: sand, silt, or clay. As shown below in Table 14.1, some subdivisions are also used.

Table 14.1 Soil Size Categories (Department of Agriculture)

Classification	Diameter (mm)
gravel	Above 2.00
very coarse sand	2.00–1.00
coarse sand	1.00–0.50
medium sand	0.50–0.25
fine sand	0.25–0.10
very fine sand	0.10–0.05
silt	0.05–0.002
clay	below 0.002

Sand. Sand grains are large enough to be seen and can be felt when rubbed between the fingers. Because of the large grain size and uneven surfaces, sand particles make only limited contact with other surfaces. Therefore, sand is not sticky nor plastic and does not make stable aggregates in soils. The pore spaces in sand are very large and infiltration capacity and permeability to water and air is high.

Silt. Silt particles cannot be seen nor felt when rubbed between the fingers. Silt is more cohesive and adhesive than sand but has only limited stickiness. Silt has much smaller pore spaces than sand, and therefore, has lower permeability than sand. Like sand, however, it contributes little to aggregate formation in soils.

Clay. Clay particles are so small they can only be seen using an electron microscope. Extensive contact between clay particles compared to sand and silt results in stickiness and plasticity. Also, clay forms hard clods when dry. Pore spaces in clay are very small and therefore infiltration capacity and permeability is extremely low.

Soils are grouped into 12 **textural classes** that are shown in Figure 14.2. The textural classes identify the particle sizes that determine the properties of a soil. **Loams** are soils that have intermediate properties.

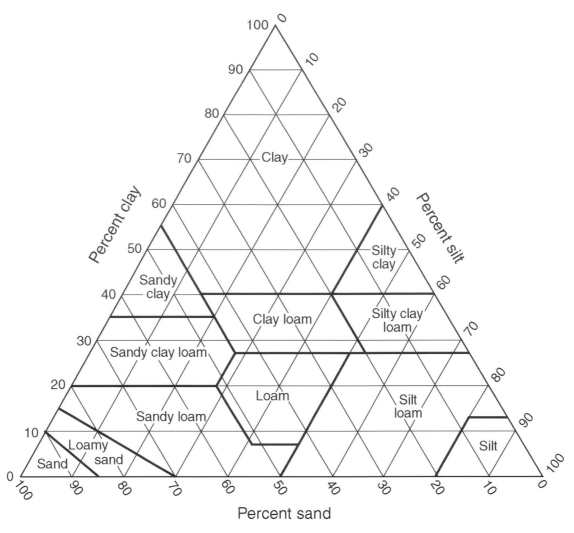

Figure 14.2   Soil textural classes. (U.S. Department of Agriculture.)

Determine Soil Texture

1. A rough approximation of soil texture can be made by using a graduated tube and timer. This method uses the same principle as standard scientifically more accurate methods: the rate of settling of soil particles in water. Soil samples will be collected or provided by the lab instructor.

    a. Fill about one third of graduated cylinders with each of the soil samples.

    b. Shake or tap the soil down so that there are no large air spaces and record the number of cubic centimeters:

	Volume		Weight	
sample 1:	_____	cc	_____	g
sample 2:	_____	cc	_____	g
sample 3:	_____	cc	_____	g
sample 4:	_____	cc	_____	g

c. In each of the cylinders add about 2 drops of Soil Dispersing Reagent for every 3 cubic centimeters of soil. Then, fill the cylinders with water to within about 5–6 cm of the top. Close or cap the top and shake vigorously for 2 minutes. Make sure the soil and water are thoroughly mixed.

d. Place the cylinder in a rack for exactly 30 seconds. *Be sure not to disturb the cylinder*. Then, measure the depth of the deposited material and record these values on the table below.

Let the solution stand for 30 minutes, then measure the depth of the deposited material again, and record these values on the same table. The first reading is the depth of the sand and the second is the depth of the sand and silt combined. The remaining particles are clay. The amount of clay can be determined by subtracting the total of sand and silt from total volume (cc).

	First Reading (cc) (sand)	Second Reading (cc) (silt and sand)	Remaining Particles (cc) (clay)
sample 1	_____	_____	_____
sample 2	_____	_____	_____
sample 3	_____	_____	_____
sample 4	_____	_____	_____

e. Determine the percentage of sand, silt, and clay for each sample by comparing the amount of each with the amount of the original soil. Refer to the Guide for the Soil Textural Classification (Figure 14.2) to determine the textural type of each sample.

	Percentage			Textural Classification
	sand	silt	clay	
sample 1	____	____	____	_____
sample 2	____	____	____	_____
sample 3	____	____	____	_____
sample 4	____	____	____	_____

f. Which of the samples has the largest pore spaces? _____

g. Which of the samples has: highest infiltration capacity? _____

highest permeability? _____

lowest infiltration capacity? _____

lowest permeability? _____

Explain:

h. Which of the samples has: highest water holding capacity? _____

lowest water holding capacity? _____

Explain:

2. Soil texture can be estimated by feel. Sand is coarse and gritty, silt has a smooth feel, and wet clay is sticky and somewhat plastic. The smallest particles you can see are coarse silt. Based on feel, estimate the textural class of soil samples provided by the lab instructor.

First, feel the soil when it is dry. Hard soil samples, those that do not easily crumble, contain at least a moderate amount of clay. Take a small soil sample and add water so that it has the consistency of modeling clay. Try to form a ribbon or wire by rolling the soil between your fingers. If a long thin wire can be easily formed, the soil is plastic and probably contains at least 40 percent clay. If a long ribbon can be formed but easily breaks, the soil is probably a clay loam or silty clay loam.

If the wet soil has a gritty feel, the soil has a sandy texture. If the feel is mostly like smooth flour and not very gritty, the soil has a silty texture. Wet sand will leave your hands moist but clean, a loamy sand (soils with a higher silt content) will leave them slightly soiled.

Textural Classification

sample 1 _____

sample 2 _____

sample 3 _____

sample 4 _____

Determine Bulk Density

Bulk density is the weight of a dry soil sample divided by its volume, or:

$$\text{Bulk Density (BD)} = \frac{\text{dry soil weight}}{\text{soil volume}}$$

and is usually expressed in grams/cm^3. The greater the bulk density, the less the pore space. For example, solid rock has no pore spaces, and therefore, a high weight per volume and high bulk density. In contrast, rock fragments will contain numerous pore spaces because they do not fit perfectly together. Therefore, the weight per volume and bulk density are comparatively less. Generally, bulk density is lowest near the soil surface and decreases with depth.

The bulk density of fine textured soils generally range from 1.0 to 1.3 g/cm^3 and for sandy soils from about 1.4 to 1.7 g/cm^3. Tillage loosens soils and lowers bulk density. Compaction will raise bulk density.

Undisturbed soil samples will be collected from the field or provided by the lab instructor. *These samples should not be compacted nor broken apart.* Metal ring samplers are used by soil scientists to remove undisturbed soil cores. If these are not available, you can use a shallow tin can with both ends removed, a block of wood, mallet, and soil sample bag.

a.   Use a shovel or knife to smooth the area to be sampled.

b.   Place the tin can on the soil surface and put the block of wood on top. Use the mallet to drive the can squarely into the soil, being sure to avoid compaction. Dig out the buried can and trim the excess soil from the bottom, again being careful to avoid compaction.

c.   Remove the samples from the cans or sample bags and allow them to dry thoroughly.

d.   Record the volume (cubic centimeters) and weight (after drying) of each sample on the table below. Calculate the bulk density by dividing the weight of the sample by its volume.

	Volume (cc)	Weight (g)	Bulk Density (weight/volume)
sample 1	_____	_____	_____
sample 2	_____	_____	_____
sample 3	_____	_____	_____
sample 4	_____	_____	_____

Particle Size and Surface Area

Chemical activity occurs on soil surfaces and therefore the amount of chemical activity is proportional to the amount of the surface area. Finer soils (those consisting of smaller particle sizes) have a greater total surface area. For example, a solid rock has a low surface:area ratio. If you crush the rock into many pieces, the total surface area for all the pieces is much higher, even though the volume has not changed. Therefore, the surface:area ratio is much higher. The same is true of sand that weathers into finer particles (silt or clay); the surface:area ratio of the soil increases.

1.   How much surface area is exposed on a cube:

      1-cm on a side: _____ cm^2

2.   What would be the total surface area if the 1-cm cube were divided into cubes:

	cm^2	m^2	particle size class
.01 cm on a side?	_____	_____	_____
.001 cm on a side?	_____	_____	_____
.0001cm on a side?	_____	_____	_____

      Refer to Table 14.1 to determine the particle size class.

3.   Which particle size class has the greatest chemical activity?

      _____

Soil Profile Data Sheet 1

Location: _____

Climate: _____

Vegetation: _____

Parent Material: _____

Topography: _____

Time (Developmental Stage): _____

Discernible horizon properties:

O: _____

_____

_____

A: _____

_____

_____

B: _____

_____

_____

C: _____

_____

_____

Soil Survey Report:

Series: _____

Order: _____

Capability Class: _____

Remarks:

Soil Profile Data Sheet 2

Location: _____

Climate: _____

Vegetation: _____

Parent Material: _____

Topography: _____

Time (Developmental Stage): _____

Discernible horizon properties:

O: _____
_____
_____

A: _____
_____
_____

B: _____
_____
_____

C: _____
_____
_____

Soil Survey Report:

Series: _____

Order: _____

Capability Class: _____

Remarks:

Soil Properties: Internet Resources

http://www.nrcs.usda.gov/ The Natural Resources Conservation Service (NRCS) is part of the U.S. Department of Agriculture (USDA). The **Maps, Facts, and Figures** section of this site provides soil maps and information on dominant soil orders.

Select **Soils and Soil Types**.

What dominant soil orders occur in your location?

Lat. _____ ; Long. _____

1.   Soil Order: _____

Description:

2.   Soil Order: _____

Description:

3.   Soil Order: _____

Description:

http://www.statlab.iastate.edu/cgi-bin/osd/osdname.cgi This is a database for official USDA-NRCS soil series descriptions. Soil series represent the most detailed level of classification in the U.S. Soil Taxonomy system.

Enter the name of local soil series.

1.   _____

General description:

Topographic characteristics of areas with this soil:

Climate conditions of areas with this soil:

Drainage and permeability:

Landuse and vegetation:

Other states where it occurs:

2. _____

General description:

Topographic characteristics of areas with this soil:

Climate conditions of areas with this soil:

Drainage and permeability:

Landuse and vegetation:

Other states where it occurs:

Soil Erosion

Equipment

> distance tapes
> stadia rod
> soil sampler
> soil sample bags

Soil erosion caused by runoff has been a major problem in the U.S. It can result in declining agricultural productivity and adversely affect water quality in nearby streams. Water erosion occurs naturally on most landscapes. However, erosion has greatly accelerated during the past century because of construction and urban development, off-road vehicles, deforestation, and a variety of agricultural practices.

In this exercise you will estimate the rate of erosion from a designated site and then determine how erosion will likely increase or decrease depending on the type of land-use changes that occur.

The **Fort Mitchell** topographic map and corresponding aerial photograph (Figure 15.1) shows an area in central Alabama with a variety of land uses. The lab instructor may instead provide a map or photograph depicting an area in your location. Select the boundaries of a site (problem area) on the map or photograph for which you will calculate the rate of erosion.

Soil erosion on the designated site will be estimated using the **universal soil-loss equation**. This soil-loss equation includes the 4 principal factors that affect runoff erosion: precipitation, type of soil, topography, and vegetation cover. This equation estimates soil erosion in tons per acre per year (A):

$$A = R * K * S * C$$

> where: R = rainfall index
> K = soil erodability factor
> S = slope factor, both steepness and length
> C = plant cover factor

Figure 15.1   Fort Mitchell, Alabama-Georgia.

Fort Mitchell, Alabama–Georgia

Scale   1:24,000

Contour Interval 10 Feet

Estimate of Soil Loss

1. Select the **rainfall erosion index** for the study site.

    High intensity rainfall causes high rates of soil erosion. The incidence of high intensity storms combined with the total annual precipitation of an area were used by the Soil Conservation Service to develop a **rainfall erosion index**.

    Select the rainfall erosion index for the study area from Figure 15.2 and enter below.

    Rainfall Erosion Index (R) = _____

2. Assign a **soil erodability factor** (K).

    This factor (K) indicates the susceptibility of the soil to sheet and rill erosion by water and is based on several soil properties (silt/fine sand and organic matter content, structure, permeability).

    The factor (K) values range from 0.05 to 0.69. The higher the value the more susceptible the soil is to water erosion. These values can be obtained from the Soil Conservation Service and are usually listed in County Soil Surveys. If K is not available, use Table 15.1 to select a value based on soil texture:

Figure 15.2   Rainfall erosion index map.

Table 15.1   Soil Erodability Factor (K)

clay or clay loam	0.2
silt-clay loam	0.3
silt loam	0.3
sandy loam	0.25
gravelly loam	0.2

Soil Erodability Factor (K) = _____

3.   Determine the **slope factor (S)**.

The ability of rainfall to erode a surface increases with an increase in either the length or gradient of the slope. Both of these factors directly affect the velocity of flow downslope. Based on the topographic map, determine the slope factor to be used for the study site.

a.   Calculate the percent grade: _____ percent

b.   Determine the length of the slope: _____ feet

Using these slope characteristics refer to Table 15.2 to determine the slope factor (S).

Slope Factor (S) = _____

Table 15.2   Slope Factor Based on Steepness and Length

Slope length (feet)	Slope Steepness (percent)								
	4	6	8	10	12	14	16	18	20
100	0.4	0.7	1.0	1.4	1.8	2.3	2.8	3.4	4.2
200	0.6	0.9	1.4	1.9	2.6	3.3	4.1	4.8	5.9
300	0.7	1.2	1.7	2.4	3.1	4.0	5.0	5.9	7.2
400	0.8	1.3	2.0	2.7	3.6	4.6	5.7	6.8	8.3
500	0.9	1.5	2.2	3.1	4.0	5.2	6.4	7.6	9.3
600	1.0	1.6	2.4	3.3	4.4	5.7	7.0	8.3	10.2
700	1.1	1.8	2.6	3.6	4.8	6.1	7.6	9.0	11.1
800	1.2	1.9	2.8	3.8	5.1	6.5	8.1	9.6	11.8
1000+	1.3	2.1	3.1	4.3	5.7	7.3	9.1	10.8	13.2

4.   Determine the **plant cover factor (C)**.

Plant cover provides significant protection from soil erosion. Foliage intercepts rainfall, reducing the impact of raindrops and slowing the rate at which water reaches the ground surface. Also, plant roots can effectively hold soil particles in place.

Use the aerial photograph to estimate:

a. Canopy cover (percentage of the site covered by trees or shrubs):

_____ percent

b. Ground cover (percentage of the site covered by grass and other low lying plants):

_____ percent

Using these vegetation cover values, refer to Table 15.3 to determine the plant cover factor (C).

Plant Cover Factor (C) = _____

Table 15.3   Plant Cover Factors

| | | Percent Groundcover | | | | |
		0	20	40	60	80	100
	0	.45	.20	.10	.042	.013	.003
		.45	.20	.15	.090	.043	.011
	25	.39	.18	.09	.039	.013	.003
Percent		.39	.22	.14	.085	.042	.011
Canopy							
Cover	50	.39	.16	.08	.038	.012	.003
		.39	.19	.13	.080	.040	.011
	75	.27	.10	.08	.035	.012	.003
		.32	.18	.12	.080	.040	.011

Use the top value if the ground surface is covered with grass or organic material at least 2 inches deep. Use the bottom value if the surface is sparsely covered with weeds and there is little organic matter.

Multiply the soil erodability factor (K), slope factor (S), and the plant cover factor (C), times the Rainfall Index (R) to determine soil loss (A):

A = (R) _____ x (K) _____ x (S) _____ x (C) _____

The soil loss on this site: _____ tons/acre/year

Predict Future Soil Loss

The previous calculation of soil loss was based on conditions that occurred when the photograph was taken. The rate of erosion, however, can be changed by altering slope steepness/length, and vegetation cover, or by changes in land-use.

1. Assume the site is left undisturbed so that a forest stand develops. As the canopy develops it shades the understory and the ground cover is reduced. Predict soil-loss on this site assuming:

   canopy cover _____ percent

   ground cover _____ percent

   In this case, the soil erodability factor (K) and slope factor (S) do not change.

   The soil loss on this site: _____ tons/acre/year

   Compared to the previous problem, the (increase, decrease) in soil-loss is:

   _____ tons/acre/year

   _____ percent

   Explain why changes in vegetation will affect the rate of soil loss as shown in this problem.

2. Assume the site is cleared for cultivation or construction. In this case, canopy cover and ground cover values are 0 percent. The soil erodability factor (K) and slope factor (S) do not change.

   The soil loss on this site: _____ tons/acre/year

   Compared to the first problem, the (increase, decrease) in soil-loss is:

   _____ tons/acre/year

   _____ percent

   Explain why soil erosion is higher when vegetation is removed as shown in this problem.

Estimate Soil Loss

Your lab instructor will select a study site for you to visit to calculate soil loss.

1. Use Figure 15.2 to determine the rainfall erosion index (R) = _____

2. Assign a soil erodability factor (K). These values can be obtained from the County Soil Survey or may be provided by your lab instructor. If K is not available, use Table 15.1 to select a value based on soil texture.

    Soil Erodability Factor (K) = _____

3. Determine the slope factor (S)

    a. Calculate the percent grade: _____ percent

    b. Determine the length of the slope: _____ feet

    Using these slope characteristics refer to Table 15.2 to determine the slope factor (S). To determine percent grade and length of the slope, use equipment provided by your lab instructor. If no equipment is available, you can make fairly reliable estimates by pacing and sighting on points upslope.

    Slope Factor (S) = _____

4. Determine the plant cover factor (C)

    Visually estimate:

    a. canopy cover (percentage of the site covered by trees and shrubs):

    _____ percent

    b. ground cover (percentage of the site covered by grass and low lying plants):

    _____ percent

    Refer to Table 15.3 to determine the plant cover factor (C).

    Plant Cover Factor (C) = _____

    Soil loss (A) = (R) _____ x (K) _____ x (S) _____ x (C) _____

    The soil loss on this site : _____ tons/acre/year

Gully Development

Soil erosion is a normal and expected condition on almost all natural surfaces. However, the rates of erosion are usually small unless a site has been disturbed by removal of vegetation or changes in hydrologic conditions. If water is channeled, its velocity increases and gully development is likely to occur. The purpose of this exercise is to determine the amount and rate of soil erosion in a gully.

Your lab instructor will select an area in which water has been channelized and as a result a gully has developed.

Calculate the Volume of the Gully

1. Divide the gully into a few sections as shown in Figure 15.3. The boundaries between each section should be _____ ft. apart (equidistant). At each line separating sections measure (in feet):

    a. the width at the top and bottom of the gully

    b. the depth (at least three measures of depth that are equidistant apart)

d	Depth
L	Length
wT	Width across top
wB	Width across bottom

Figure 15.3 Calculate the volume of the gully.

192

Enter these values below:

Section 1:   width top _____   depth _____
                   width bottom _____   depth _____
                                              depth _____
                                            depth _____

Section 2:   width top _____   depth _____
                   width bottom _____   depth _____
                                              depth _____
                                            depth _____

Section 3:   width top _____   depth _____
                   width bottom _____   depth _____
                                              depth _____
                                            depth _____

Section 4:   width top _____   depth _____
                   width bottom _____   depth _____
                                              depth _____
                                            depth _____

2.   Determine the cross-sectional area for each section boundary.

Find the *average* for the depth measurements (feet) and multiply by the average of the widths (feet)(top and bottom).

Section 1:

depth _____ x _____ width
=cross-sectional area (cs1) _____ feet^2

Section 2:

depth _____ x _____ width
=cross-sectional area (cs2) _____ feet^2

Section 3:

depth _____ x _____ width
=cross-sectional area (cs3) _____ feet^2

Section 4:

depth _____ x _____ width
=cross-sectional area (cs4) _____ feet^2

3. Determine the volume of the gully:

Average the cross-sectional area (cs) between consecutive section boundaries and multiply by the length between each to determine the volume of that section of the gully.

volume of the first gully segment = _____ feet^3

volume of the second gully segment = _____ feet^3

volume of the third gully segment = _____ feet^3

volume of the fourth gully segment = _____ feet^3

Add the volumes of the gully segments to determine the total volume of soil loss.

_____ feet^3

Determine the Weight of Soil Loss

The class will be divided into groups for soil sample collection. Extract an undisturbed soil sample that will fit in a graduated cylinder. Be sure not to compact the soil sample or allow it to break apart. Record its volume:

_____ cm^3 (ml)

Dry the soil to determine its weight: _____ g

Convert the weight of this soil sample to lbs/feet^3, using the following conversions:

1 kg (1000g) = 2.204 lbs
1 cm^3 (ml) = 0.061 inches^3

(dry weight) _____ lbs/(volume) _____ feet^3

If you were unable to obtain a soil sample to determine its weight, use the following values:

clay   80 lbs/ft^3
clay or silt loams   100 lbs/ft^3
sand or sandy loams   120 lbs/ft^3

Determine the total weight of the soil lost from the gully. (1 ton = 2000 lbs)

_____ lbs/feet^3 x _____ total soil loss (feet^3)

soil loss: _____ tons

Years since the gully formed: _____

Average soil loss per year: _____ tons

The average rate of soil loss on cultivated land in the U.S. is about 4 tons/acre/year. Soil loss greater than about 5 tons/acre/year is considered serious and could result in declining agricultural productivity. How does the rate of erosion for the gully you measured compare to that of the U.S.?

Soil Erosion: Internet Resource

http://www.nrcs.usda.gov/ The Natural Resources Conservation Service is part of the U.S. Department of Agriculture. The *Maps, Facts, and Figures* section of this site provides information on soil and water conservation, agricultural productivity, and changing land-use patterns.

Select *Soil Erosion*.

What areas of the U.S. have excessive cropland soil erosion caused by water?

Are there areas near your location that have excessive cropland soil erosion by water or by wind? Where?

Note: The regions of *Prime Farmland* in the U.S. Which of these regions have excessive rates of soil erosion?

Vegetation Analysis

Equipment

distance measuring tapes
tree diameter measuring tapes
increment borer
survey stakes or stake wire flagging

Vegetation has a strong influence on the environment by modifying microclimate, soil properties, and hydrology, and provides an important control on soil erosion and slope stability. It is necessary to consider vegetation for a full understanding of the physical environment. In this exercise you will use common ecological methods to describe forest communities, interpret forest vegetation structure, and determine growth rates for different tree species.

Describing Forest Communities

Determining density and dominance of individual tree species is the most widely used method to describe forest communities. These are simple techniques to quantitatively rank species within a stand.

Density is the number of individual plants per unit area and is calculated by a simple count of plants on a designated site. For example, Species A has a density of 23 individuals per $1000m^2$ (study site area) and Species B has 4 individuals on the same site. This is a simple method. But direct comparisons of density give information about the reproductive ability of a species on a specific site. Stand sizes used by different groups may not be the same size. To compare densities from different sites, densities can be converted to a standard size area ($1000m^2$, $10,000m^2$, etc.).

Relative density is the density of one species as a percent of the total density, which is calculated by adding density values for all plants on the study site. Table 16.1 details an example of density and relative density in central Georgia.

Table 16.1   Middle Slope Position in Central Georgia

Species	Density (number)	Relative Density (percent)
Loblolly Pine (Pinus taeda)	15	6
Chestnut Oak (Quercus prinus)	85	34
White Oak (Quercus alba)	55	22
Flowering Dogwood (Cornus Florida)	95	38
Total	250	100 percent

Note: Common names are not standard for plant species. Always include scientific names for identification to avoid confusion.

Table 16.1 shows that a total of 250 individual trees were recorded on a 1000m^2 forest stand. Fifteen loblolly pine occurred on the stand. This was 6 percent relative density. Density described the number of individuals present, but tells us nothing about the size or dominance of these species.

When describing forest communities, **dominance** is usually determined by measuring **basal area**, which is the cross-sectional area of the tree trunk approximately 1.3 meters above the ground. Basal area is expressed as m^3/ha. Dominance for each species in the stand is determined by measuring basal area for each tree of that species and then adding the values.

Relative dominance is the dominance of one species as a percent to the total dominance in a stand. Dominance values from the same site are shown in Table 16.2.

Table 16.2   Dominance Values from Central Georgia

Species	Dominance (m^2) (total basal area)	Relative Dominance (percent)
Loblolly Pine	.35	70
Chestnut Oak	.10	20
White Oak	.04	8
Flowering Dogwood	.01	2
Total	.50	100 percent

Loblolly pine had lowest density, but each individual was very large. Therefore, loblolly, based on basal area, was the dominant species. Conversely, flowering dogwood is a small species, rarely growing more than a few cm in diameter. Flowering dogwood had the most individuals on the site, but had a relative dominance of only 2 percent.

Determine Density and Dominance

Figures 16.1 and 16.2 are maps of forest stands showing the location and diameter of the two most dominant species on each site. Both sites are 40 x 30 meters and are divided into 5 x 5 meter subplots.

1. Calculate density and relative density and enter these values on the table below.

2. Calculate dominance using basal area. Worksheets are provided at the end of this exercise. The diameter of each tree is shown on the map. Using the diameter, basal area can be easily calculated using the following: basal area = 3.14 * radius^2. Conversion values are in Appendix B.

Baldcypress—Green Ash Stand (Figure 16.1)

Species	Density	Rel. Den.(%)	Dominance(m^2)	Rel. Dom.(%)
Baldcypress	_____	_____	_____	_____
Green Ash	_____	_____	_____	_____

Loblolly Pine—Sweetgum Stand (Figure 16.2)

Species	Density	Rel. Den.(%)	Dominance(m^2)	Rel. Dom.(%)
Loblolly Pine	_____	_____	_____	_____
Sweetgum	_____	_____	_____	_____

Bald Cypress (Taxodium distichum)

Green Ash (Fraxinus Pennsylvanian)

Figure 16.1 Map of a forest stand.

◉ Loblolly Pine (Pinus taeda)

■ Sweetgum (Liquidambar styraciflua)

Figure 16.2 Map of a forest stand.

Size Structure Analysis

Size structure analysis is a common technique used to determine how the density and dominance of species in a forest stand are likely to change in the future. In Table 16.3 below are size-class distributions for 3 tree species found on the same study site.

Table 16.3   Size-Class Distributions for 3 Species

Species	Size-Classes (cm diameter)					
	<2	2–10	11–20	21–30	31–40	41–50
A	220	81	47	22	10	2
B	312	45	2	0	0	0
C	0	0	0	0	18	6

Species A has individuals in all size-classes. There are many small individuals and progressively smaller numbers in the larger size-classes. This indicates that Species A has been continuously regenerating on this site since the oldest first established. Not all the individuals in one size-class will survive and grow into the next size-class. So we expect progressively smaller numbers in the larger size-classes.

Species B is only represented in smaller size-classes. Possibly, this species does not grow very large, but is successfully reproducing in the understory of Species A. It is also possible Species B will grow very large, but in this case its initial establishment occurred much later than Species A.

Species C has only large individuals that are about the same size. These individuals probably established about the same time. The absence of small individuals indicates this species is not regenerating. Unless conditions change, it is unlikely this species will replace itself on this site. This size-class distribution is common among shade intolerant species that establish after forest stands are destroyed by fire, hurricanes, ice storms, etc.

Note: You do not know the age of these individuals. In most cases, the very large trees will be older and those in much smaller size-classes. But you cannot be certain a larger individual is older than a smaller one unless you age the trees.

Determine Size-Class Distribution

Count the number of individuals in each size-class for both species shown in Figures 16.1 and 16.2. Enter these values on the tables below.

Baldcypress—Green Ash Stand (Figure 16.1)

Species	Size-Classes (cm diameter)						
	<2	2–10	11–20	21–30	31–40	41–50	+50
Baldcypress	____	____	____	____	____	____	____
Green ash	____	____	____	____	____	____	____

1.   Baldcypress: diameter Min: _____, Max: _____

2. Describe regeneration status for baldcypress.

3. Green ash: diameter Min: _____, Max: _____

4. Describe the regeneration status for green ash.

Loblolly Pine—Sweetgum Stand (Figure 16.2)

Species	Size-Classes (cm diameter)						
	<2	2–10	11–20	21–30	31–40	41–50	+50
Loblolly pine	_____	_____	_____	_____	_____	_____	_____
Sweetgum	_____	_____	_____	_____	_____	_____	_____

1. Loblolly pine: diameter Min: _____, Max: _____

2. Describe regeneration status for loblolly pine.

3. Sweetgum: diameter Min: _____, Max: _____

4. Describe the regeneration status for sweetgum.

Spatial Distribution

The distribution of a species within a forest stand may be random, meaning that an individual has the same probability of occurring anywhere in the stand. In many cases, however, a species is clustered. The clustering, or grouping of individuals of a species can be caused by (1) more *favorable environmental conditions* on certain parts of the site increasing the probability of regeneration there. This may include more favorable sunlight or soil moisture conditions. (2) Reproductive characteristics of the species may cause clustering. For example, some tree species reproduce by producing root sprouts. In other cases, seeds are very heavy and stay near the mature trees after they fall.

Refer to Figure 16.1.

Green ash spatial distribution: random: _____ or clustered _____

If clustered, what are the likely causes for this spatial pattern?

Refer to Figure 16.2.

Sweetgum spatial distribution: random: _____ or clustered _____

If clustered, what are the likely causes for this spatial pattern?

Describe Forest Stands

The lab instructor will designate a forest stand(s) and indicate species you will describe and evaluate. Use blank forest stand maps at the end of this exercise to plot the position of each individual of the species you are describing. Use sample maps of forest stands (Figures 16.3 and 16.4) if field work is not possible.

1. Establish the study site boundary using distance measuring tapes. Survey stakes or flags can be used to mark the boundary and distances along the boundary depending on the size of the subplots.

2. Use distance measuring tapes to determine the positions of the subplots.

3. Plot the position of each individual for each species designated by the lab instructor.

4. Measure the diameter of each tree approximately 1.3m above the ground and record this value on the map.

5. Calculate density and enter these values on the **Density and Dominance Worksheet**.

6. Use the **Basal Area Worksheets** provided at the end of this exercise to calculate **Total Basal Area**. Then enter this value on **Density and Dominance Worksheet**.

● Species: _____

■ Species: _____

Figure 16.3   Map of forest stand.

◎ Species: _____

■ Species: _____

Figure 16.4 Map of forest stand.

Determine Size-Class Distribution of Forest Stands

For the forest stands you describe above (or stands shown in the sample maps) evaluate the size-class distribution.

Study Site 1 (or Figure _____)

Species	Size-Classes (cm diameter)						
	<2	2–10	11–20	21–30	31–40	41–50	+50
1. _____	____	____	____	____	____	____	____
2. _____	____	____	____	____	____	____	____

Species

1. _____ diameter: Min. _____, Max._____
2. _____ diameter: Min. _____, Max._____

Describe the regeneration status for each species.

Study Site 2 (or Figure _____)

Species	Size-Classes (cm diameter)						
	<2	2–10	11–20	21–30	31–40	41–50	+50
1. _____	____	____	____	____	____	____	____
2. _____	____	____	____	____	____	____	____

Species

1. _____ diameter: Min. _____, Max._____
2. _____ diameter: Min. _____, Max._____

Describe the regeneration status for each species.

Evaluate Spatial Patterns of Species

For the forest stands you describe above (or stands shown in the sample maps) evaluate the spatial patterns for each species.

Study Site 1 (or Figure _____)

       Species   Spatial distribution

1. _____   random: _____ or clustered _____
2. _____   random: _____ or clustered _____

If any species are clustered, explain the likely causes for this spatial pattern based on your field observations.

Study Site 2 (or Figure _____)

       Species   Spatial distribution

1. _____   random: _____ or clustered _____
2. _____   random: _____ or clustered _____

If any species are clustered, explain the likely causes for this spatial pattern based on your field observations.

Tree Age and Rates of Growth

Generally, larger trees are older than smaller ones. But this is not always the case. Some species grow much faster than others. Also, some individuals of the same species grow faster than others because of genetic or environmental factors. Trees can be aged by using an increment borer to extract a core from the tree and then counting the tree's annual rings.

    The lab instructor will select which species will be aged. Several trees from one or more of the dominant species will be aged.

1.   For each species, subjectively select individuals to represent different sizes. An increment borer will be used to extract cores so that annual rings can be counted. Core the tree near the ground (less than 30 cm height if possible) so that the ring counts will be similar

to the age of the tree. Screw in the increment borer aiming at the center of the tree. Be sure the borer enters far enough into the tree so you reach the center.

2. Extract the cores and store them in soda straws. Seal the ends of the straws with masking tape. After sealing the cores, split open the sides of the straws so the cores will dry out. Otherwise they might rot.

3. Record the diameter of all trees that are cored and label the straw so that later you can match the size (diameter of the tree) to age that will be determined by counting the annual rings.

4. In the lab, remove the cores from the straws and count the annual rings to determine the age of the tree. Enter the diameter and matching ring counts (age of the tree) on the **Tree Age Worksheet**.

Determine Surface Age

Trees can sometimes be used to determine the approximate age of new surfaces created by the formation of river deltas, the filling of oxbow lakes and development of point-bars or mid-channel islands in alluvial streams, and mass movement that exposes previously underlying material. On many of these sites, trees quickly establish. The age of the oldest trees can be used to determine the approximate age, or at least the minimum age, of the surface.

The lab instructor will select sites for you to determine the minimum age. At each site, select the trees that appear to be the oldest based on size. Use an increment borer to extract cores and store them in straws as described above.

In the lab, remove the cores from the straws and count the annual rings to determine the age of the tree. Enter the ring counts (age of the tree) **Surface Age Worksheet**.

Density and Dominance Worksheet

Study Site 1

Study site location: _____

Slope position: _____

Slope aspect: _____

Site size:_____ x _____ m, Subplot size:_____ x _____ m

Species	Density	Rel. Den.(%)	Dominance(m^2)	Rel. Dom.(%)
1. _____	_____	_____	_____	_____
2. _____	_____	_____	_____	_____
3. _____	_____	_____	_____	_____
4. _____	_____	_____	_____	_____
5. _____	_____	_____	_____	_____
6. _____	_____	_____	_____	_____
Total	_____	_____ (%)	_____ (m^2)	_____ (%)

Study Site 2

Study site location: _____

Slope position: _____

Slope aspect: _____

Site size:_____ x _____ m, Subplot size:_____ x _____ m

Species	Density	Rel. Den.(%)	Dominance(m^2)	Rel. Dom.(%)
1. _____	_____	_____	_____	_____
2. _____	_____	_____	_____	_____
3. _____	_____	_____	_____	_____
4. _____	_____	_____	_____	_____
5. _____	_____	_____	_____	_____
6. _____	_____	_____	_____	_____
Total	_____	_____ (%)	_____ (m^2)	_____ (%)

Basal Area Worksheet #1

Site number:_____ Species: _____

Diameter	BA(m^2)	Diameter	BA(m^2)	Diameter	BA(m^2)
1. _____ _____		27. _____ _____		54. _____ _____	
2. _____ _____		28. _____ _____		55. _____ _____	
3. _____ _____		29. _____ _____		56. _____ _____	
4. _____ _____		30. _____ _____		57. _____ _____	
5. _____ _____		31. _____ _____		58. _____ _____	
6. _____ _____		32. _____ _____		59. _____ _____	
7. _____ _____		33. _____ _____		60. _____ _____	
8. _____ _____		34. _____ _____		61. _____ _____	
9. _____ _____		35. _____ _____		62. _____ _____	
10. _____ _____		36. _____ _____		63. _____ _____	
11. _____ _____		37. _____ _____		64. _____ _____	
12. _____ _____		38. _____ _____		65. _____ _____	
13. _____ _____		39. _____ _____		66. _____ _____	
14. _____ _____		40. _____ _____		67. _____ _____	
15. _____ _____		41. _____ _____		68. _____ _____	
16. _____ _____		42. _____ _____		69. _____ _____	
17. _____ _____		43. _____ _____		70. _____ _____	
18. _____ _____		44. _____ _____		71. _____ _____	
19. _____ _____		45. _____ _____		72. _____ _____	
20. _____ _____		46. _____ _____		73. _____ _____	
21. _____ _____		47. _____ _____		74. _____ _____	
22. _____ _____		48. _____ _____		75. _____ _____	
23. _____ _____		49. _____ _____		76. _____ _____	
24. _____ _____		50. _____ _____		77. _____ _____	
25. _____ _____		51. _____ _____		78. _____ _____	
26. _____ _____		52. _____ _____		79. _____ _____	
		53. _____ _____		80. _____ _____	

Total Basal Area: _____ m^2

Basal Area Worksheet #2

Site number:_____ Species: _____

Diameter	BA(m^2)	Diameter	BA(m^2)	Diameter	BA(m^2)
1. _____	_____	27. _____	_____	54. _____	_____
2. _____	_____	28. _____	_____	55. _____	_____
3. _____	_____	29. _____	_____	56. _____	_____
4. _____	_____	30. _____	_____	57. _____	_____
5. _____	_____	31. _____	_____	58. _____	_____
6. _____	_____	32. _____	_____	59. _____	_____
7. _____	_____	33. _____	_____	60. _____	_____
8. _____	_____	34. _____	_____	61. _____	_____
9. _____	_____	35. _____	_____	62. _____	_____
10. _____	_____	36. _____	_____	63. _____	_____
11. _____	_____	37. _____	_____	64. _____	_____
12. _____	_____	38. _____	_____	65. _____	_____
13. _____	_____	39. _____	_____	66. _____	_____
14. _____	_____	40. _____	_____	67. _____	_____
15. _____	_____	41. _____	_____	68. _____	_____
16. _____	_____	42. _____	_____	69. _____	_____
17. _____	_____	43. _____	_____	70. _____	_____
18. _____	_____	44. _____	_____	71. _____	_____
19. _____	_____	45. _____	_____	72. _____	_____
20. _____	_____	46. _____	_____	73. _____	_____
21. _____	_____	47. _____	_____	74. _____	_____
22. _____	_____	48. _____	_____	75. _____	_____
23. _____	_____	49. _____	_____	76. _____	_____
24. _____	_____	50. _____	_____	77. _____	_____
25. _____	_____	51. _____	_____	78. _____	_____
26. _____	_____	52. _____	_____	79. _____	_____
		53. _____	_____	80. _____	_____

Total Basal Area: _____ m^2

Basal Area Worksheet #3

Site number:_____ Species: _____

Diameter	BA(m^2)	Diameter	BA(m^2)	Diameter	BA(m^2)
1. _____	_____	27. _____	_____	54. _____	_____
2. _____	_____	28. _____	_____	55. _____	_____
3. _____	_____	29. _____	_____	56. _____	_____
4. _____	_____	30. _____	_____	57. _____	_____
5. _____	_____	31. _____	_____	58. _____	_____
6. _____	_____	32. _____	_____	59. _____	_____
7. _____	_____	33. _____	_____	60. _____	_____
8. _____	_____	34. _____	_____	61. _____	_____
9. _____	_____	35. _____	_____	62. _____	_____
10. _____	_____	36. _____	_____	63. _____	_____
11. _____	_____	37. _____	_____	64. _____	_____
12. _____	_____	38. _____	_____	65. _____	_____
13. _____	_____	39. _____	_____	66. _____	_____
14. _____	_____	40. _____	_____	67. _____	_____
15. _____	_____	41. _____	_____	68. _____	_____
16. _____	_____	42. _____	_____	69. _____	_____
17. _____	_____	43. _____	_____	70. _____	_____
18. _____	_____	44. _____	_____	71. _____	_____
19. _____	_____	45. _____	_____	72. _____	_____
20. _____	_____	46. _____	_____	73. _____	_____
21. _____	_____	47. _____	_____	74. _____	_____
22. _____	_____	48. _____	_____	75. _____	_____
23. _____	_____	49. _____	_____	76. _____	_____
24. _____	_____	50. _____	_____	77. _____	_____
25. _____	_____	51. _____	_____	78. _____	_____
26. _____	_____	52. _____	_____	79. _____	_____
		53. _____	_____	80. _____	_____

Total Basal Area: _____ m^2

Basal Area Worksheet #4

Site number:_____ Species: _____

Diameter	BA(m²)	Diameter	BA(m²)	Diameter	BA(m²)
1. _____	_____	27. _____	_____	54. _____	_____
2. _____	_____	28. _____	_____	55. _____	_____
3. _____	_____	29. _____	_____	56. _____	_____
4. _____	_____	30. _____	_____	57. _____	_____
5. _____	_____	31. _____	_____	58. _____	_____
6. _____	_____	32. _____	_____	59. _____	_____
7. _____	_____	33. _____	_____	60. _____	_____
8. _____	_____	34. _____	_____	61. _____	_____
9. _____	_____	35. _____	_____	62. _____	_____
10. _____	_____	36. _____	_____	63. _____	_____
11. _____	_____	37. _____	_____	64. _____	_____
12. _____	_____	38. _____	_____	65. _____	_____
13. _____	_____	39. _____	_____	66. _____	_____
14. _____	_____	40. _____	_____	67. _____	_____
15. _____	_____	41. _____	_____	68. _____	_____
16. _____	_____	42. _____	_____	69. _____	_____
17. _____	_____	43. _____	_____	70. _____	_____
18. _____	_____	44. _____	_____	71. _____	_____
19. _____	_____	45. _____	_____	72. _____	_____
20. _____	_____	46. _____	_____	73. _____	_____
21. _____	_____	47. _____	_____	74. _____	_____
22. _____	_____	48. _____	_____	75. _____	_____
23. _____	_____	49. _____	_____	76. _____	_____
24. _____	_____	50. _____	_____	77. _____	_____
25. _____	_____	51. _____	_____	78. _____	_____
26. _____	_____	52. _____	_____	79. _____	_____
		53. _____	_____	80. _____	_____

Total Basal Area: _____ m²

Tree Age Worksheet

Species: 1. _____   2. _____

	Diameter (cm)	Age (yr)	Growth Rate (cm/yr)	Diameter (cm)	Age (yr)	Growth Rate (cm/yr)
1.	_____	_____	_____	_____	_____	_____
2.	_____	_____	_____	_____	_____	_____
3.	_____	_____	_____	_____	_____	_____
4.	_____	_____	_____	_____	_____	_____
5.	_____	_____	_____	_____	_____	_____
6.	_____	_____	_____	_____	_____	_____
7.	_____	_____	_____	_____	_____	_____
8.	_____	_____	_____	_____	_____	_____
9.	_____	_____	_____	_____	_____	_____
10.	_____	_____	_____	_____	_____	_____

Diameter: min. _____, max. _____   min. _____, max. _____

Age: min: _____, max. _____   min: _____, max. _____

Average growth rate: _____ cm/yr   _____ cm/yr

Surface Age Worksheet

Surface description:

    Site 1. _____

    Site 2. _____

    Site 3. _____

	Site 1	Site 2	Site 3
Species:	_____	_____	_____

Annual rings (years)

	Site 1	Site 2	Site 3
1.	_____	_____	_____
2.	_____	_____	_____
3.	_____	_____	_____
4.	_____	_____	_____
5.	_____	_____	_____

Maximum age:

    _____   _____   _____

Year of initial establishment = current year – max. age):

    _____   _____   _____

Species 1. _____   3. _____

        2. _____   4. _____

Figure 16.5 Map of forest stand.

Species 1. _____   3. _____

2. _____   4. _____

Figure 16.6   Map of forest stand.

LAB EXERCISE

17

Runoff and Infiltration

Most precipitation will either infiltrate the soil surface or runoff, and only a small proportion will evaporate. The purpose of this exercise is (1) to determine for a specific area the amount of rainfall that runs-off and eventually enters streams, or that infiltrates the ground surface, and (2) to determine how human impact can alter the runoff/infiltration ratio.

Water can readily infiltrate most dry soils. As water enters the soil surface the pore spaces in the soils become filled. The rate of infiltration slows and much of the additional precipitation will runoff. On naturally vegetated surfaces there is little runoff except during very heavy periods of precipitation. In contrast, paved surfaces and those occupied by buildings will have no infiltration, and therefore, all precipitation runs-off.

Figure 17.1 is an aerial photograph showing the suburban edge of Huntsville, Alabama. The built-up areas are mostly residential. The lab instructor will select a portion of this photograph or may provide a photograph showing an area in your location for which you will determine the runoff/infiltration ratio.

The R.F. of the photograph is 1: _____.

First, calculate the designated area: _____ m²
A standard area measurement is a hectare (ha) which is 100x100 meters, or 10,000m². Convert the designated area to hectares.

_____ ha

Next, estimate the percentage of the drainage basin that is paved. Typically, about one-half of the surface occupied by single family housing is paved. The percentage of the surface paved is usually much higher for other types of land-use (commercial, industrial).

Percentage of the surface that is paved: _____

No infiltration occurs on paved surfaces, and therefore, runoff is 100 percent of precipitation.

On vegetated or non-paved surfaces precipitation will infiltrate the ground surface, but the amount will vary depending in large part on the amount of water already in the soil and different soil types. If there is very little precipitation no runoff will occur. However, after water enters the soil surface the infiltration rate slows because many of the pore spaces in the surface soil layer are now filled with water.

The infiltration rate also depends on soil texture. Sandy soils have large pore spaces that allow water to move easily through them. The finer texture soils (silt and clay) have smaller pore sizes that restrict water movement, and therefore, infiltration is slower.

Figure 17.1   Suburban edge of Huntsville, Alabama.

Use the following table (17.1) to determine rates of infiltration for different soil textural classes during (1) the first 30 minutes of precipitation, and (2) for the remaining period.

Table 17.1   Rate of Infiltration on Vegetated (Non-Paved) Surfaces

Soil Textural Class	Rate of Infiltration (cm/hour)	
	first 30 mins.	remaining period
clay or clay loam	0.4	0.1
silt-clay loam	1.0	0.2
silt loam	1.2	0.3
sandy loam	2.0	0.4

Soil Textural Class _____

Infiltration   a.   first 30 mins. _____ (cm/hour)

                  b.   remaining period _____ (cm/hour)

Suppose 6 cm of rain falls on the designated area in a 24 hour period.

1.   What is the average rainfall intensity in cm per hour?

_____

2.   Assuming this intensity to hold true during the first 30 minutes, how much runoff would occur?

from the paved area _____ cm

from the vegetated area _____ cm

3.   After 30 minutes how much runoff would occur per hour?

from the paved area _____ cm

from the vegetated area _____ cm

4.   How much runoff would occur in the entire 24 hour period?

from the paved area _____ cm

from the vegetated area _____ cm

Use the following equation to determine the volume of runoff (V) in cubic meters:

$$V = (A * R)/100$$

where: A = the area (m^2)
R = the runoff (cm)

Note: The above value (A * R) is divided by 100. Remember, there are 100 cm in a meter, and this allows V to be measured in cubic meters.

5. How many cubic meters ran off this site in the 24 hour period?

_____

6. What percentage of the total rain that fell ran off?

_____ percent

The runoff/infiltration ratio for the same area changes as land-use changes. For this problem assume 6 cm of rain falls on the same site in a 24 hour period, as in previous problem. However, more buildings and parking lots are constructed. Therefore, the total paved area increases and vegetated area decreases.

Assume that the paved area increases _____ percent, or _____ m^2.

What is the paved area? _____ m^2

What percentage of the drainage basin is now paved? _____ (= paved area/designated area)

vegetated? _____

1. What is the average rainfall intensity in cm per hour?

_____

2. Assuming this intensity to hold true during the first 30 minutes, how much runoff would occur?

from the paved area _____ cm

from the vegetated area _____ cm

3. After 30 minutes how much runoff would occur per hour?

from the paved area _____ cm

from the vegetated area _____ cm

4. How much runoff would occur in the entire 24 hour period?

from the paved area _____ cm

from the vegetated area _____ cm

5. How many cubic meters would runoff this site in the 24 hour period?

_____

6. What percentage of the total rain that fell ran off?

_____ percent

7. What is the percent change in runoff compared to that in the first problem in this exercise?

_____ percent (increase, decrease)

For the following problem assume _____ cm of rain falls in a _____ hour period.

Designated area_____ m²

_____ ha

Percentage of the drainage basin that is paved _____

paved area: _____ m²

vegetated area: _____ m²

Soil Textural Class _____

Infiltration   a. first 30 mins. _____ (cm/hour)

b. remaining period _____ (cm/hour)

1. What is the average rainfall intensity in cm per hour?

_____

2. Assuming this intensity to hold true during the first 30 minutes, how much runoff would occur?

from the paved area _____ cm

from the vegetated area _____ _____ cm

3. After 30 minutes how much runoff would occur per hour?

from the paved area _____ cm

from the vegetated area _____ cm

4. How much runoff would occur in the entire 24 hour period?

from the paved area _____ cm

from the vegetated area _____ cm

5. How many cubic meters would runoff this site in the 24 hour period?

_____

6. What percentage of the total rain that fell ran off?

_____ percent

7. What is the percent change in runoff compared to that in the first problem in this exercise?

_____ percent (increase, decrease)

Runoff and Infiltration: Internet Resources

Changing Landuse

http://www.nrcs.usda.gov/ The Natural Resources Conservation Service is part of the U.S. Department of Agriculture. The *Maps, Facts, and Figures* section of this site provides information on agriculture, land cover, and changing land-use patterns.

Select the sections on *Prime Farmland* and *Urbanization*.

What areas of the U.S. have the largest percentage of land that is urbanized?

What areas of the U.S. have the highest annual rate of land development?

What areas of the U.S. have recently had the most prime farmland lost to development? Why has farmland been lost faster in this region than other parts of the U.S.?

http://edcwww.cr.usgs.gov/earthshots/slow/tableofcontentsmap This site from the U.S. Geological Survey has aerial photographs taken at different times to show how man is changing the environment.

Your lab instructor will select **cities** for you to view. Based on the repeat photographs and accompanying text, how have these sites changed during the past few decades?

City: _____   Lat. _____ ; Long. _____

Changes:

City: _____   Lat. _____ ; Long. _____

Dates of Photographs:
1st: _____
2nd: _____
3rd: _____

Changes:

City: _____   Lat. _____ ; Long. _____

Dates of Photographs:
1st: _____
2nd: _____
3rd: _____

Changes:

How will the expansion of these cities affect runoff and infiltration in watersheds that were formerly naturally vegetated?

Water Quality

http://www.epa.gov/surf2/locate/ This site is from the U.S. Environmental Protection Agency. It allows you to locate watersheds and provides information about water quality. You can locate watersheds using state, county, or the river names. You can select an **Index of Watershed Indicators** (IWI) for an assessment of the watershed health that uses a scale of 1(high quality)–6 (low).

Determine the water quality for the following river's watersheds:

1. _____ State: _____ County: _____

Number of rivers and streams in the watershed: _____

Total river miles (or km): _____

Watershed area: _____

IWI score: _____   Water quality: high or low

2. _____ State: _____ County: _____

Number of rivers and streams in the watershed: _____

Total river miles (or km): _____

Watershed area: _____

IWI score: _____   Water quality: high or low

3. _____ State: _____ County: _____

Number of rivers and streams in the watershed: _____

Total river miles (or km): _____

Watershed area: _____

IWI score: _____   Water quality: high or low

Stream Discharge

Stream discharge is the amount of water flowing through a channel. Discharge directly affects velocity which determines a stream's ability to erode its channel and transport sediment. In this exercise you will (1) observe a small stream to calculate its discharge, (2) estimate velocity and discharge for streams that cannot be directly measured, and (3) predict average discharge based on the size of the drainage area and precipitation.

Calculate Discharge

Your lab instructor will select a small stream to be used in this exercise. Discharge (Q) is calculated by:

$$Q = AV$$

where: A = cross-sectional area of the channel, and
V = average velocity.

Figure 18.1   Cross-section (A), perimeter (P) and slope (S) of a stream channel.

1.   Determine the cross-sectional area of the channel.

Measure the width of the channel.

width = _____ m

Measure the water depth. If the depth varies significantly, you should make several measurements and calculate the average.

depth = _____ m

Calculate the cross-sectional area by multiplying width x water depth.

cross-sectional area = _____ m^2

2. Determine the stream velocity.

Calculate the average velocity, which is expressed as m/second. The velocity can be determined by timing the passage of a float along a known distance. For example, if your float takes 10 seconds to go 20 meters, then 20m/10 = 2 m/second.

Repeat the velocity measurement 5 times. Complete the chart below with your observations:

Observations

	1	2	3	4	5
Distance (m)	_____	_____	_____	_____	_____
Time (sec)	_____	_____	_____	_____	_____
Velocity (m/sec)	_____	_____	_____	_____	_____

Average velocity = _____ m/sec

Now, multiply A (the cross-sectional area) x V (average velocity) to calculate Q (discharge).

Q = _____ m^3/second

Based on this discharge, what is the daily volume of water that flows through this stream?

_____ m^3

Determine the Erosion Potential

The velocity of water determines its ability to move or transport sediment. Based on the stream velocity you measured, determine the largest particle size that can be moved by referring to Figure 18.2 shown below:

Maximum particle size that can be transported _____ mm

Textural class (clay, silt, etc.) _____

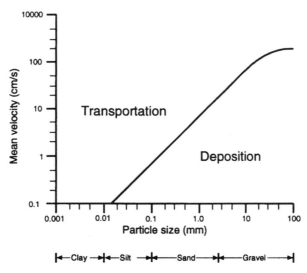

Figure 18.2   Relationship between water velocity and particle size that can be transported.

Estimation of Bankfull Discharge

Bankfull refers to the channel being completely filled with water. Any additional water will cause flooding because it cannot be held in the channel and is forced into the surrounding floodplain. Bankfull discharge can be easily calculated.

Draw a cross-section of the stream channel as accurately as possible on Figure 18.3. Then, determine the cross-sectional area for the channel (assuming the channel is completely full).

Figure 18.3   Cross-section of stream channel.

Bankfull A (cross-sectional area) = _____ m²

You must know the water velocity to estimate the bankfull discharge. In this case you do not know the velocity, because as Q increases, so does V. But in this problem assume that V = _____ m/sec.

1. Based on the channel cross-sectional area and the estimated velocity, the Q (discharge) at bankfull is

    _____ m³/sec

2. Based on this discharge, what is the daily volume of water that flows through this stream if the stage is bankfull?

    _____ m³

3. What is the percentage increase of this value compared to observed discharge?

    _____ percent

4. Based on the estimated stream velocity at bankfull, what is the largest particle size that can be moved? (Refer to Figure 18.2.)

    _____ mm

    Textural class (clay, silt, etc.): _____

Estimating Stream Velocity

The velocity of a stream can be estimated with reasonable accuracy when direct measurements are not possible. Velocity is controlled by the 3 following channel characteristics:

1. hydraulic radius (which is a measure of the proportion of water in a stream that comes in contact with the channel bed or banks)

2. channel slope or gradient

3. roughness of the channel

Stream velocity (V) can be estimated using Manning's formula which takes into account these 3 channel characteristics:

$$V = \frac{1.0}{n} * R^{.67} * S^{.5}$$

where: V =   in m/second
       R =   hydraulic radius (the cross-sectional area of the channel divided by the wetted perimeter of the channel)
       S =   channel slope or gradient
       n =   roughness coefficient

The lab instructor will select a stream shown on a topographic map. You will estimate stream velocity at (1) average flow and (2) bankfull, using Manning's formula. A worksheet to help calculate velocity is included.

The topographic map _____ has an R.F. of 1: _____

Estimate velocity for the following stream: _____
during (1) average flow and (2) at bankfull.

Enter the values on the blank lines and on the worksheet that follows.

1.   Calculate the Hydraulic Radius (R)

Larger, deeper streams generally flow faster than smaller streams that have the same gradient. The greater the discharge (and deeper the channel) the smaller the proportion of water in the channel that comes in contact with the channel banks, and is therefore slowed by friction. The hydraulic radius (R) is a value representing the ratio of (1) the cross-section of the channel (as shown in Figure 18.1) and (2) the wetted perimeter, or length of the channel bed plus the height of the banks occupied by water.

Use the topographic map to determine the width of the channel. You cannot determine the water depth from a topographic map. In this problem assume that the average water depth at normal flow is _____ percent of the channel width.

depth at normal flow: _____

Based on the depth at average flow (that is depicted on the topographic map) and the elevation of the floodplain adjacent to the channel, determine:

depth at bankfull: _____

Draw the cross-section of the stream at average flow and bankfull below on Figure 18.4.

Figure 18.4   Channel cross-section of the stream at average flow and bankfull.

To calculate the cross-sectional area determine:

a. channel width: _____ m

    channel width (bkf): _____ m

b. channel depth: _____ m

    channel depth (bkf): _____ m

Calculate the channel cross-sectional area (w * d):

    average flow: _____ m^2

    bankfull: _____ m^2

Calculate the wetted perimeter (w + 2d):

    average flow: _____ m

    bankfull: _____ m

Hydraulic radius (R) (cross-sectional area/wetted perimeter):

    average flow: _____

    bankfull: _____

2. Calculate Stream Gradient (S)

Remember that the stream gradient is a ratio of the elevational difference between 2 points along the stream and the channel length (or distance) between them. The steeper the gradient, the faster the water will move.

$$S = \frac{rise}{run}$$

rise = the elevational difference between 2 points on the channel bed

    _____ ft or m

run = the distance between these points.

    _____ ft or m

Slope or stream gradient (S): _____

3. Determine Roughness Coefficient (n)

Water moves fastest through a smooth, straight channel. Irregular channels (those that meander and have channel obstructions) have a much slower velocity. Channel roughness is a measure of channel characteristics that directly affect velocity including: channel configuration (straight, meandering, etc.), the presence of vegetation or large boulders, and if the channel sides and bottom are smooth or irregular. As channel roughness increases, flow velocity decreases.

Approximate values for n are:

Alluvial stream:
straight, clean channels	0.030
straight, weedy or with boulders	0.035
clean winding channels	0.040

Mountain streams:
cobbles and boulders	0.050
gravel bottom	0.030
sand bottom	0.020

Roughness coefficient (n) _____

WORKSHEET

	Normal Flow	Bankfull
width:	_____	_____
depth:	_____	_____
cross-sectional area:	_____	_____
wetted perimeter:	_____	_____
Hydraulic radius (R):	_____	_____
Stream gradient (S):	_____	_____
Roughness coefficient (n):	_____	_____

Enter the values for the hydraulic radius (R), stream gradient (S), and the roughness coefficient (n) into Manning's formula to estimate stream velocity.

normal flow		bankfull	
velocity: _____	m/sec	_____	m/sec
discharge: _____	m^3/sec	_____	m^3/sec

Based on this discharge, calculate the daily volume of water that flows through this stream.

daily discharge: _____ m^3 _____ m^3

Determine the largest particle size that can be moved. (Refer to Figure 18.2.)

particle size: _____ mm   _____ mm

textural class: _____   _____

1. If your calculations are correct, stream velocity is greater at bankfull than normal (average) flow. Why does the water move faster when the water level is higher?

2. Based on the relationship between velocity and the particle size that can be transported, what are the effects of floods compared to average flows on erosion and moving channel sediment?

Predicting Stream Discharge

In the previous problems in this exercise you have estimated discharge by directly measuring channel dimensions and velocity, and later by using topographic maps to indirectly determine these same stream conditions. The **average discharge** of a stream can be roughly predicted by using the following equation that takes into account the area of the drainage basin and precipitation:

$$Q = 0.082 * A^{1.02} * P^{0.75}$$

where: Q =   mean annual discharge (ft^3/second)
         A =   basin area (square miles)
         P =   mean annual precipitation (inches) minus 30

The **Heidelberg** topographic map depicts an area in central Kentucky. Predict the mean annual discharge for Cave Branch, a small stream that is a tributary of the Kentucky River. The lab instructor may instead provide a map or aerial photograph of a site closer to your location to determine discharge.

Mean annual precipitation in this drainage basin is _____ inches.

1. Determine the area of the drainage basin: _____ miles^2

2. Use these values in the above equation to predict the mean annual discharge.
   _____ ft^3/sec

3. Based on predicted discharge, what is the daily volume of water that flows through this stream?
   _____ ft^3

Heidelberg, Kentucky

Scale   1:24,000

Contour Interval 20 Feet

4. The Cave Branch watershed is mostly forested and only a small portion of its surface area has been cleared. The use of this equation to determine average discharge does not take into account land-use patterns. How will discharge be affected if a large section of the drainage that is currently covered by natural vegetation is cleared for agricultural uses, either pasture or cultivation?

5. How will discharge be affected if a large section of the drainage that is currently covered by natural vegetation is paved?

Stream Discharge: Internet Resources

River Stage and Discharge

http://www.usgs.gov/ The U.S. Geological Survey has a site with current stream stage and discharge data for rivers in every state. Select **Real-Time Water Data** to access this site. Note the map showing stream gaging stations across the U.S.

What areas of the U.S. currently have above average streamflow?

What areas have below average streamflow?

Click on this map to select sites near your location. For each site note its location, average flow (discharge), current flow, and stage.

	Station Name	average flow (cfs)	current flow (cfs)	stage (f)
Site 1	_____	_____	_____	_____
Site 2	_____	_____	_____	_____
Site 3	_____	_____	_____	_____
Site 4	_____	_____	_____	_____

Are the current flows at these sites consistently above or below average flows?

Would you expect these flows to be above or below average flows during this time of year? Explain the basis for your answer.

River Forecasting

http://www.nws.noaa.gov/ The National Weather Service has River Forecasting Centers located in different regions of the country. Select the ***Central***, ***Eastern***, or ***Southern Regions*** and then ***River Forecasting Centers*** for information on current river stage, precipitation and stage forecasts, and historical data. Note that river stage (water surface elevation) is usually measured in feet and discharge is usually measured in ft^3/second.

Select the gaging stations nearest your location.

1. River: _____   Gaging Station: _____

Lat. _____; Long. _____

Current river stage: _____   Flood stage: _____

Current discharge: _____

2. River: _____   Gaging Station: _____

Lat. _____ _____; Long. _____

Current river stage: _____   Flood stage: _____

Current discharge: _____

3. River: _____   Gaging Station: _____

Lat. _____; Long. _____

Current river stage: _____   Flood stage: _____

Current discharge: _____

http://www.riverwatch.noaa.gov/main.html This NOAA site gives stage data for many sites on the Mississippi River and its major tributaries. It is updated daily. Flood stage and five day forecasts showing changes in water surface elevation are also given.

Flood Satellite Images

http://edcwww2.cr.usgs.gov/moberly.html This site by the U.S. Geological Survey has a series of Landsat satellite images of the Missouri River floodplain taken before and after the 1993 flood. You can see the extent of flooding by viewing ***Preflood*** and ***Peak Flood*** images.

What is the dominant landuse within this portion of the Missouri River floodplain?

By looking at the preflood image, how can you determine the floodplain boundary?

What portion of the floodplain was underwater during the peakflood?

Note the image showing **_Levees and High Risk Areas_**.

Approximately what portion of the floodplain was protected by levees?

What parts of the levee system are most likely to fail during floods?

http://www.geo.mtu.edu/rs/tm/flood/ This site shows Landsat Satellite images of the upper Mississippi River flood during the summer of 1993. You can compare images taken during normal flow to those taken during the height of flooding.

LAB EXERCISE

19

Recurrence Intervals

The frequency of natural events (such as flooding, droughts, and natural fires) can be estimated by using historical records. Rather than predicting that a natural event will occur in a given year (e.g., the Mississippi River will flood in the year 2004), the technique for calculating a recurrence interval is an attempt to determine, or at least estimate the probability that an event of a certain magnitude will occur in any given year. The purpose of this exercise is to explain the method for estimating the probability of the magnitude and frequency of natural events and to recognize the limitations of this technique.

The recurrence interval is determined by:

$$RI = (n + 1) / m$$

where: RI = recurrence interval in years
        n   = number of years of observations
        m   = rank of the observation

The following table is a 13 year record (1978–1991) of annual precipitation for a climatic recording station in Colorado. Each year has been ranked so that the wettest is #1 and the driest is #13. The recurrence interval (in years) has been calculated using the above equation in an attempt to estimate the probability of this area receiving a certain amount of precipitation for any one year.

Year	Precip (in)	Rank	Recurrence Interval (yrs)
1978	22.87	5	2.8
1979	12.08	11	1.3
1981	14.83	9	1.5
1982	13.88	10	1.4
1983	19.75	7	2.0
1984	16.02	8	1.7
1985	22.68	6	2.3
1986	9.78	13	1.0
1987	30.07	1	14.0
1988	27.49	2	7.0
1989	27.31	3	4.6
1990	25.09	4	3.5
1991	10.75	12	1.1

So, in this example it has been estimated that a year with at least 25 inches of precipitation (1990) will occur once every 3.5 years. A year with at least 30.0 inches of annual precipitation (1987) will occur once every 14 years.

The amount of annual precipitation and corresponding return intervals are plotted on Figure 19.1. A line has been drawn that best fits these points.

Predictions of the frequency (recurrence interval) of a higher annual precipitation are based on continuing the line with the same slope beyond the data until it intersects the vertical axis on the right-hand side of the graph.

Based on the precipitation curve in Figure 19.1, how much annual precipitation will occur at least once every:

50 years? _____ inches

100 years? _____ inches

200 years? _____ inches

In this case only 13 years of data was used. A longer record will provide for more accurate predictions.

Probability

Based on the recurrence interval, you can estimate the probability that an event of a certain magnitude will occur in any one year. The probability (P) is the inverse of the recurrence interval for any one year, and is calculated by:

$$P = 1 / RI * 100$$

Therefore, for the year 1990 the return interval is 3.5 years (for there to be at least 25 inches of precipitation). Therefore:

$$1 / 3.5 * 100 = 28.5$$

or, a 28.5% probability of at least that much rainfall (or an event of that magnitude) occurring for any one year.

Precipitation Recurrence Interval

A record of the highest annual rainfall events for each year between 1970 and 1989 for Baton Rouge, Louisiana, is shown below. Determine the rank and recurrence intervals (RI) for the data set. The highest rainfall year will be ranked #1 and the lowest will be ranked #20.

Precipitation Curve

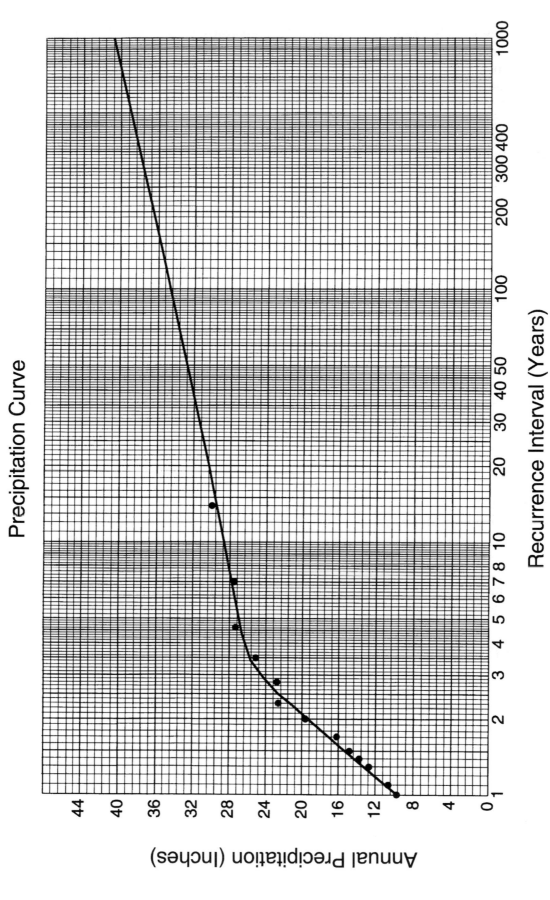

Annual Precipitation (Inches)

Recurrence Interval (Years)

Figure 19.1 Precipitation curve.

Year	Maximum Rainfall	Rank	RI
1970	2.69	_____	_____
1971	3.28	_____	_____
1972	3.85	_____	_____
1973	5.13	_____	_____
1974	2.02	_____	_____
1975	3.08	_____	_____
1976	2.61	_____	_____
1977	4.51	_____	_____
1978	2.59	_____	_____
1979	4.07	_____	_____
1980	6.37	_____	_____
1981	2.29	_____	_____
1982	7.25	_____	_____
1983	7.02	_____	_____
1984	4.59	_____	_____
1985	2.57	_____	_____
1986	3.20	_____	_____
1987	7.31	_____	_____
1988	3.32	_____	_____
1989	7.81	_____	_____

Plot each of the data points on Figure 19.2.

Draw a line that best fits the data to show the overall trend. As in Figure 19.1, the line will not necessarily be straight. Although the data is for only 20 years, you can determine the 50 year (and 100 or 500 year) maximum annual rainfall events by continuing the line that fits the data (with the same slope) until it intersects the vertical axis on the right-hand side of the graph. Note that this line is not horizontal but is inclined upward.

Based on Figure 19.2 answer the following questions.

1. What is the probability that for any one year there will be at least one rainfall event no less than:

> 6 inches? _____ %
>
> 7 inches? _____ %
>
> 8 inches? _____ %

2. What is the predicted   50 year rainfall event? _____ inches

> 100 year rainfall event? _____ inches
>
> 500 year rainfall event? _____ inches

Precipitation Curve

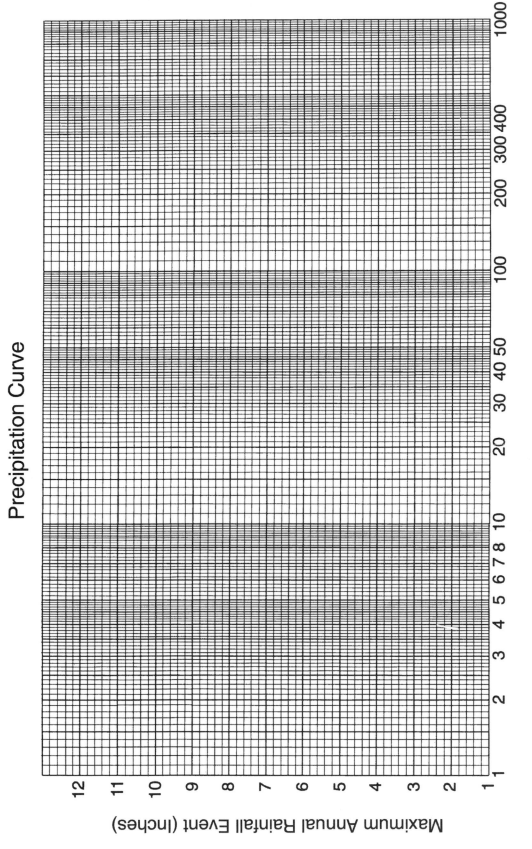

Recurrence Interval (Years)

Maximum Annual Rainfall Event (Inches)

Figure 19.2 Precipitation curve.

3. Will the predicted 50, 100, and 500 year rainfall events necessarily occur at those time intervals? Explain.

4. Will future rainfall patterns and events reflect what has happened in the recent past? Explain.

Flood Recurrence Interval

The following table is a yearly record for the highest stream **stage** from the Hatchie River in western Tennessee. Stage is the height of the water, or the **water surface elevation** above sea level. This data can be used to determine how often a stream floods and what portion of a floodplain will be submerged.

Determine the rank and recurrence intervals (RI) for the data set. The highest stage will be ranked #1.

Year	Maximum Stage	Rank	RI
1956	256.6	_____	_____
1957	257.9	_____	_____
1958	254.8	_____	_____
1959	254.9	_____	_____
1960	254.7	_____	_____
1961	256.1	_____	_____
1962	257.2	_____	_____
1963	253.5	_____	_____
1964	255.2	_____	_____
1965	258.2	_____	_____
1966	255.3	_____	_____
1967	255.0	_____	_____
1968	255.6	_____	_____
1969	257.8	_____	_____
1970	255.4	_____	_____
1971	255.5	_____	_____
1972	255.7	_____	_____
1973	259.5	_____	_____
1974	256.3	_____	_____

Plot each of the data points on Figure 19.3.

Draw a line that best fits the data to show the overall trend, and continue the line with the same slope until it intersects the vertical axis on the right-hand side of the graph.

Based on Figure 19.3 answer the following questions.

Flood Frequency Curve

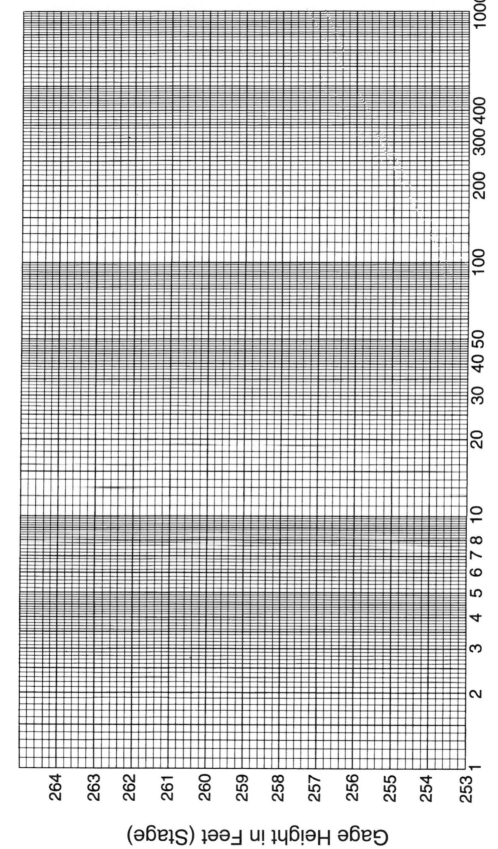

Recurrence Interval (Years)

Gage Height in Feet (Stage)

Figure 19.3   Flood frequency curve.

253

1. Flood stage is the elevation at which water is above the channel banks and therefore moving across the floodplain. Flood stage is 253 feet at this gaging station. Based on the stage record, what is the probability that for any year the Hatchie River will flood?

   _____ %

2. What is the probability for any year that stage will reach a height of 256 feet above sea level?

   _____ %

3. What is the probability for any year that stage will reach a height of 259 feet above sea level?

   _____ %

4. What is the predicted stage for the 50 year flood?

   _____ feet elevation

   Remember, you do not have 50 years of data. The 50 year flood is only a prediction. The farther you go beyond the limits of your data (100, 200, 500 year floods) the less reliable the prediction.

5. What is the predicted stage for the 100 year flood?

   _____ feet elevation

   The differences in yearly maximum stage may seem small. But, floodplains are areas of low relief, and slight increases in water level can cause large areas near the river channel to go under water.

   The **South Ripley** map shows a section of the Hatchie River in western Tennessee. The stage data in the table above was collected at the gaging station on the bridge crossing the river. You can determine approximately what portion of the floodplain will be submerged during a flood by matching stage data with the elevation on the map which is shown by contour lines. Contour line intervals on this map are 10 feet (and 5 feet in some areas of the floodplain).

6. Measure along the highway on the north side of the river to determine how far from the river channel the floodplain will be covered with water if stage is:

   255 feet: _____

   260 feet: _____

7. How far from the river channel will the floodplain be covered with water during the 50 year flood?

   _____

Ripley South, Tennessee

1		$\frac{1}{2}$		0				1 MILE

1000	0	1000	2000	3000	4000	5000	6000	7000 FEET

1	.5	0	1 KILOMETER

Scale   1:24,000

Contour Interval 10 Feet

8. Approximately what portion of the floodplain will be submerged during the 50 year flood? (This can be determined by calculating the distance floodwater extends from the active channel to the edge of the floodplain.)

_____

9. The Garden of Everlasting Life cemetery is located on the south side of the river. At what stage will flooding cover the cemetery?

_____ feet

Will water reach this cemetery during a 100-year flood? _____

10. Calculate the probability that a levee built to hold a 100 year flood will be over-topped by flood-water during

any one year: _____

a 10 year period: _____

a 50 year period: _____

11. The size of flood control structures such as dams and levees are usually based on the 100 (or 500) year flood. Do you regard this as a reliable method to determine the height of flood-control levees? Explain.

A Natural Event

Historical data of a natural event will be provided by the lab instructor. The event is:

_____

Record the data on the following table and calculate the recurrence interval. After calculating the return interval, plot the data on Figure 19.4 and answer the questions that follow.

Year	Event	Rank	RI
_____	_____	_____	_____
_____	_____	_____	_____
_____	_____	_____	_____
_____	_____	_____	_____
_____	_____	_____	_____
_____	_____	_____	_____
_____	_____	_____	_____
_____	_____	_____	_____
_____	_____	_____	_____
_____	_____	_____	_____
_____	_____	_____	_____
_____	_____	_____	_____
_____	_____	_____	_____
_____	_____	_____	_____
_____	_____	_____	_____
_____	_____	_____	_____
_____	_____	_____	_____
_____	_____	_____	_____
_____	_____	_____	_____
_____	_____	_____	_____
_____	_____	_____	_____
_____	_____	_____	_____
_____	_____	_____	_____
_____	_____	_____	_____
_____	_____	_____	_____

Based on Figure 19.4 answer the following questions.

1. What is the magnitude of the predicted

   50 year event? _____

   100 year event? _____

   200 year event? _____

2. What is the probability that for any one year there will be at least one event no less (or greater) than:

   (magnitude) _____ : _____ %

   (magnitude) _____ : _____ %

   (magnitude) _____ : _____ %

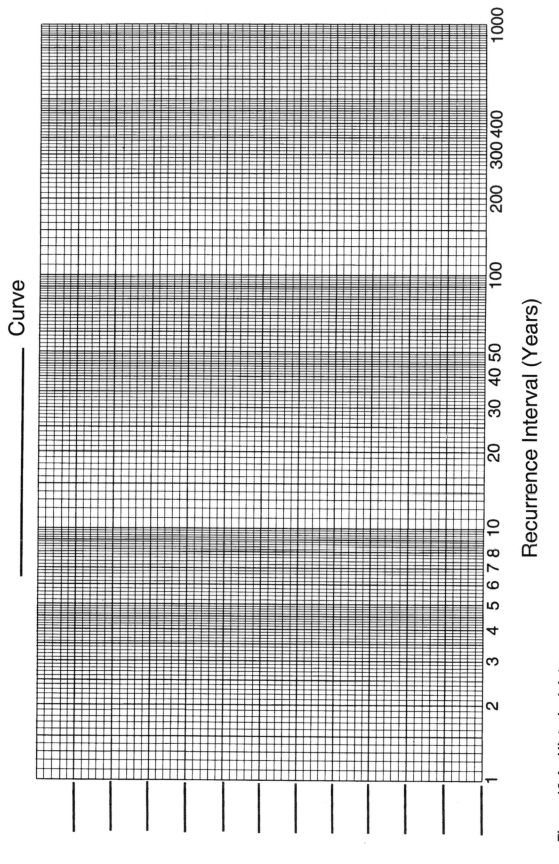

Figure 19.4   Historical data.

259

LAB EXERCISE

20

Aerial Photograph Interpretation and Landform Analysis

Aerial photographs are among the most useful tools for physical geographers. They can be used for mapping, city and regional planning, forest inventories, landscape evaluation, studies in earth sciences, and have many other purposes. Taking photographs of the same area (repeat photographs) over a long period of time can be used to determine both the rates and the extent of environmental change.

Interpretation requires determination of the photograph scale. The scale or representative fraction (RF) of a photograph (and of any map) indicates a relationship between the distance on the ground and corresponding distance on the photograph. For example, an RF of 1:12,000 means that one inch on the map or photograph equals 12,000 inches on the ground. Other distance measures (such as cm or feet) could also be used.

Refer to the _____ topographic map and corresponding aerial photograph in this exercise.

The topographic map has an R.F. of 1: _____

Based on this scale, you can determine the R.F. of the corresponding aerial photograph. Indicate the distance units you use (circle one): *inches* or *cm*

1. Select two landmarks or coordinates that can be easily located on both the topographic map and photograph.

2. Measure the distance between the two points on the photograph: _____ (cm or inches). Enter this value under 4a.

    Determine the distance between these two points on the map: _____ (cm or inches)

3. Convert the distance on the map to the real ground distance using the above R.F.

    Distance on the ground _____ (meters or feet). Enter this value under 4b.

4. The photograph scale is calculated by determining the ratio of the following values:

a. Distance on the photo _____

b. Distance on the ground _____

    R.F. = 4a : 4b

    Convert these values to equivalent units of measure.

5. Photograph R.F. (scale) is 1: _____

6. Determine the scale (R.F.) of the aerial photographs that correspond to the following topographic maps in this exercise:

a. Topographic map _____ : map R.F. 1: _____

    air photo R.F.: 1: _____

b. Topographic map _____ : map R.F. 1: _____

    air photo R.F.: 1: _____

c. Topographic map _____ : map R.F. 1: _____

    air photo R.F.: 1:_____

d. Topographic map _____ :map R.F. 1: _____

    air photo R.F.: 1: _____

7. Determine the distance on the ground between the designated coordinates on the aerial photographs that correspond to the following topographic maps in this exercise:

a. Topographic map _____

    points: (1) _____ and (2) _____

                    _____ feet

                    _____ miles

                    _____ meters

                    _____ kilometers

b. Topographic map _____

    points: (1) _____ and (2) _____

                    _____ feet

                    _____ miles

                    _____ meters

                    _____ kilometers

8. Determine the area on the ground that corresponds to the designated area on the aerial photographs and topographic map in this exercise:

a. Topographic map _____

Scale of the corresponding aerial photograph:

1: _____

      Area on the photograph is _____ x _____ (in. or cm).

      Corresponding area on the ground:

                    _____ $feet^2$

                    _____ $mile^2$

                    _____ m^2

                    _____ km^2

b. Topographic map _____

Scale of the corresponding aerial photograph:

1: _____

      Area on the photograph is _____ x _____ (in. or cm).

      Corresponding area on the ground:

                    _____ $feet^2$

                    _____ $mile^2$

                    _____ m^2

                    _____ km^2

Stereoscopic Vision

Aerial photographs can be viewed in stereoscopic (or three-dimensional) vision. The ability to visualize the relief of a site enhances your ability to study photographs. Aerial photographs are taken from low flying aircraft. Photographs are usually taken so that they overlap. They can be viewed in stereoscopic vision by using a stereoscope to look at two different photographs at the same time.

The two pictures are placed with one overlapping the other so that the feature you want to view is about the same distance apart as your eyes (2–2½ inches). A stereoscope is placed over the photographs and allows you to look at the same spot on the two photographs simultaneously. The angular difference between the camera in the airplane and the positions on the ground when the picture was taken creates slightly different photographs which allows three-dimensional viewing, but with significant vertical exaggeration.

Refer to Figure 20.1 which is a stereogram of a steep-sided volcanic cone. This stereogram was chosen because it can easily be seen by the first time stereo viewer.

Stereoscopes will be provided by the lab instructor. Place them over the stereogram. While looking through the stereoscope, slowly adjust its position until you see a clear three-dimensional image of the landscape.

Figure 20.1 Mt. Capulin in New Mexico. (Photograph courtesy University of Illinois Committee on Aerial Photography and Remote Sensing.)

Landform Analysis

The following sections in this exercise deal with (1) **fluvial process**, (2) **arid landscapes**, and (3) **glaciated landscapes**. Topographic maps and corresponding aerial photographs are included for a visual descriptions of each section.

Fluvial Processes

Running water is a dominant factor shaping the earth's surface. The nature of most landforms can be related to stream processes. Other means of erosion, such as glaciation, waves, and wind, may dominate in some areas, but none have as significant an impact over a large area of the earth. This section describes stream processes and how these processes affect the landscape.

1. **Stream Incision** (Reference: **Cape Solitude**, Arizona, Quadrangle)

    Streams with steep gradients have a high water velocity and the ability to erode or cut down the channel bed. These are referred to as **degrading streams**. The downward cutting creates steep V-shaped valleys with very little if any flat-land in the valley bottom. Rapids and waterfalls are common on these streams.
    The **Cape Solitude** map depicts the Colorado River where it passes through Marble Canyon upstream of the Grand Canyon. These streams have dissected the Colorado Plateau creating steep canyons with vertical cliffs and high local relief. Figure 20.2 is a stereogram of a portion of this map.

    a. What is the relief of the land area shown in the:

        map? _____

        photo?_____

    b. Cape Solitude is located south of the mouth of the Little Colorado River. What is Cape Solitude's:

        coordinates on the photo? _____

        elevation? _____

    c. How deep is the canyon below the rim at Cape Solitude? _____

    d. The mouth of Sixtymile Creek is about 2.5 km north of the mouth of the Little Colorado River. What are the coordinates on the photo for this position?

        _____

    e. What is the stream gradient of Sixtymile Creek? _____

    f. A delta has formed at the mouth of the Little Colorado River. What caused its formation?

Figure 20.2   Marble Canyon in Colorado. (Photograph courtesy University of Illinois Committee on Aerial Photography and Remote Sensing.)

Cape Solitude, Arizona

Scale   1:24,000

Contour Interval 40 Feet

2. **Lateral Erosion** (Reference: **Walsh Knolls**, Utah-Colorado, Quadrangle)

As a stream erodes its channel bed, its elevation decreases. By cutting-down over long periods of time, the stream gradient becomes gentler and both water velocity and the rate of erosion slows. Eventually the elevation of the stream bed **stabilizes**. That is, the stream has reached a **state of equilibrium**, or a balance, between its ability to erode and the amount of sediment it carries. Once a stream is no longer downcutting it will begin **lateral erosion** and the creation of **meanders** that gradually widen the stream valley and create a **floodplain**.

The White River and many of its tributaries have dissected a plateau in western Colorado. The White River is now moving laterally and has created a narrow floodplain. Figure 20.3 is a stereogram of a portion of the **Walsh Knolls** map.

a. What is the relief of the land area shown in the:

map? _____

photo? _____

b. In which direction is the White River flowing? _____

c. Cottonwood Creek enters the White River in Section 24. What is the width of the White River floodplain at the mouth of Cottonwood Creek?

_____ m

d. The White River has begun to meander, indicating that it is no longer cutting down the channel bed. Are the tributaries of the White River down-cutting or moving laterally? Evidence?

e. The topographic map depicts roads and buildings. Are these or any other human features visible in the photo?

Figure 20.3 White River in Colorado. (Photograph courtesy University of Illinois Committee on Aerial Photography and Remote Sensing.)

Walsh Knolls, Utah–Colorado

Scale 1:24,000

Contour Interval 20 Feet

3. **Meandering Stream** (Reference: **Fishhook Lake**, Mississippi, Quadrangle)

Meandering streams erode the surrounding valley sides and uplands, and over long periods of time can create broad, flat valleys. The flat surfaces adjacent to the meandering channel comprise its floodplain, and at least occasionally will be covered by floodwater. Almost all surfaces within these floodplains were at one time occupied by the active stream channel. As streams move laterally the meanders often come together creating a **cutoff**. The stream then changes course by flowing through the cutoff and leaving an **abandoned meander** or **oxbow lake**. Oxbow lakes eventually fill-in with sediment becoming **meander scars**. These former channels are depicted on topographic maps and in many cases distinguishable on aerial photographs.

The **Fishhook Lake** map depicts the Tallahatchie River within a broad, nearly flat floodplain. Figure 20.4 is a stereogram of a small portion of this map. Many oxbow lakes and meander scars are visible on the photo.

a. What is the relief of the land area shown in the:

map? _____

photo? _____

b. The Tallahatchie River flows through central Mississippi, an area that normally supports forest vegetation. The Tallahatchie River floods almost every year, and yet most of the area in the floodplain is in cultivation. Why is there intensive agriculture in an area subject to flooding?

c. Black Lake (A.6-1.5) is a former river channel. Why is the width of Black Lake in the photo smaller than that of the active channel?

d. The aerial photographs (Figure 20.4) were taken in 1941. The corresponding topographic map was made from photographs taken in 1978, 37 years later. By comparing photographs and the map you can determine in many ways how the landscape was changed during this period. How have the following features changed?

(1) active channel

Figure 20.4   Horseshoe Lake in Mississippi. (Photograph courtesy University of Illinois Committee on Aerial Photography and Remote Sensing.)

Fishhook Lake, Mississippi

Scale   1:24,000

Contour Interval 5 Feet

(2) Horseshoe Lake

(3) forest vegetation

4. **Stream Rejuvenation** (Reference: **The Loop**, Utah, Quadrangle)

Streams that have created floodplains by lateral movement can renew downcutting if for any reason stream gradient increases. A steeper gradient will increase water velocity and result in a stream again eroding its channel bed. The renewed downcutting of a formerly meandering stream is defined as **rejuvenation**, which can be caused by (1) tectonic uplift, (2) decrease in base level, or (3) a change in climatic conditions that result in greater runoff. The stream will downcut to a lower elevation and may eventually create a new floodplain.

The **Loop** map depicts the Colorado River flowing through a deep canyon. The former meandering stream began to downcut, resulting in **channel incision** and the **entrenchment** of meanders. Figure 20.5 is a stereogram of a small portion of this map.

a. What is the relief of the land area shown in the:

map? _____   _____

photo? _____

b. In which direction is the Colorado River flowing? _____

c. What is the width of the meander neck (land surface between the channels) at C.9-2.4?

_____ m

d. Why is no floodplain evident along this stretch of the river?

e. Is there any evidence that the Colorado River tributaries have also entrenched? Explain.

Figure 20.5   The Loop in Utah. (Photograph courtesy University of Illinois Committee on Aerial Photography and Remote Sensing.)

THE

LOOP

COLORADO

RIVER

Pack

4WD

4WD

Salt

CANYONLANDS

The Loop, Utah

Scale   1:24,000

Contour Interval 40 Feet

5. **Braided Stream** (Reference: **Nevens**, Nebraska, Quadrangle)

In contrast to meandering streams, **braided streams** have multiple, shifting channels. Low-gradient streams typically develop this type of river pattern if they have very high sediment loads. Channels are separated by sand or gravel bars. High discharge will redistribute the sediment causing laterally shifting channels.

The **Nevens** map depicts a braided section of the North Platte River in Nebraska. This stream is flowing toward the Mississippi River to the east. It is feed by snow melt-water from the Rocky Mountains. Figure 20.6 is a stereogram of a small portion of this map.

a. Vegetation on the mid-channel bars separating channels would suggest the bars are somewhat stable. Can you distinguish if bars are vegetated or not? Explain.

b. There is a single water channel in the right photograph at the edge of the North Platte River valley. Is this a natural channel or man-made? Explain.

c. There is vegetation within the river banks. But as shown on the right photograph, there is little vegetation on the surrounding upland. Explain.

d. Is there evidence of agriculture on the photograph?

Figure 20.6   North Platte River in Nebraska. (Photograph courtesy University of Illinois Committee on Aerial Photography and Remote Sensing.)

Nevens, Nebraska

Scale   1:24,000

Contour Interval 20 Feet

Arid Landscapes

Dry regions of the world have distinctive physical features. In many places soils are either thin or absent and vegetation is sparse. Most streams are ephemeral, flowing only during or after rainfall or from melting snow in surrounding mountains. No permanent streams originate in desert or semi-desert regions. Also, deserts contain rivers that flow into basins or valleys and do not drain into any ocean. In deserts, evaporation exceeds precipitation and rivers become progressively smaller with distance downstream.

Although most desert streams are ephemeral, water is the dominant agent shaping the landscape. Vegetation provides little cover to intercept rain and the absence of surface organic matter and root systems allows erosion and rapid movement of sediment. This section describes landforms in dry climates that include desert and steppe regions.

1. **Alluvial Fans** (Reference: **Stovepipe Wells**, California, Quadrangle)

    Some streams have such a high sediment load that not all the material can be transported downstream. Instead the sediment is deposited which raises the channel bed elevation. The **alluvial fans** depicted on the **Stovepipe Wells** map have formed at the edge of Death Valley. They develop when ephemeral streams flowing in the surrounding mountain canyons emerge into the desert basin. Because of the lower stream gradient in the desert basin, water velocity slows and the sediment is deposited. Figure 20.7 is a stereogram of a small portion of this map.

a. What is the stream gradient in Mosaic Canyon? _____

b. What is the gradient of the same stream after it flows out of the mouth of the Canyon (D.1-2.5)?

    _____

c. Multiple stream channels on the alluvial fans can easily be seen on the photo. Why are there so many channels on these surfaces as opposed to only one main channel in the canyon?

d. What is the dark line near the edge of the fans (B.9-3.0)?

e. What is the nature of vegetative cover on and beyond the edge of the fans?

f. What are the photo coordinates for the town of Stovepipe Wells?

    _____

Figure 20.7   Stovepipe Wells in California. (Photograph courtesy University of Illinois Committee on Aerial Photography and Remote Sensing.)

Stovepipe Wells, California

Scale   1:24,000

Contour Interval 40 Feet

2. **Meandering Stream and Headward Erosion** (Reference: **Scenic**, South Dakota, Quadrangle)

As mentioned earlier, dry regions have sparse or no vegetation. Although rain is infrequent, erosion can be severe compared to humid regions with more dense vegetation. To contrast erosion potential in humid and dry regions, consider how forest vegetation modifies hydrology. First, the forest canopy intercepts rain. During heavy rain water will eventually begin dripping through the canopy. The surface organic layer and the upper mineral soil horizon that is mixed with organic material effectively holds water. Only after long periods of rain do the upper soils become saturated. If surface infiltration is no longer possible, water begins to move across the ground surface, potentially moving sediment. In most dry regions, water will move across the surface much more quickly, which has notable effect on the landscape, as shown in the following photographs.

The **Scenic** map depicts a small river with a tortuous meandering pattern in South Dakota. Tributaries are extending their valleys into the surrounding uplands. Figure 20.8 is a stereogram of a small portion of this map.

a. Has this river developed a floodplain? Evidence?

b. What is the ratio of the width of the floodplain to the width to the meander belt?

c. Have the smaller tributary streams developed floodplains? _____

d. What is the difference in elevation between the floodplain and surrounding plateau?

_____

e. Is there any agriculture in the area? Evidence?

f. Are any trees shown in the photograph? Where?

Figure 20.8 Bear Creek in South Dakota. (Photograph courtesy University of Illinois Committee on Aerial Photography and Remote Sensing.)

North 71 Table Sch

Scenic, South Dakota

Scale   1:24,000

Contour Interval 10 Feet

3. **Dissection of a Plateau** (Reference: **Sallies Spring**, New Mexico, Quadrangle)

As shown on the previous map and photograph, the **Sallies Spring** map depicts tributaries of a small meandering river that are incising and extending their valleys into the surrounding uplands. In this case, overlying rock of the upland plateau is highly resistant to erosion resulting in the development of cliffs. Note the shallow stream pattern at the top of the photograph. This stream is at a much higher elevation than Bear Creek, but water drains away from the river. Figure 20.9 is a stereogram of a small portion of this map.

a. What is the difference in elevation between the San Juan River floodplain and plateau near the Drill Hole south of the river?

    _____

b. What is the width of the

    meander belt of San Juan River? _____ m

    floodplain of the San Juan River? _____ m

c. Is there any evidence of sediment deposition by the San Juan River? What depositional features are present?

d. The photographs were taken in September, one of the driest months of the year in New Mexico. At the time of the photograph was water flowing through the:

    San Juan River? _____

    tributaries of the San Juan river? _____

e. What is the gradient of tributary streams before they enter the San Juan River floodplain?

Figure 20.9   San Juan River in New Mexico. (Photograph courtesy University of Illinois Committee on Aerial Photography and Remote Sensing.)

294

Sallies Spring, New Mexico

Scale   1:24,000

Contour Interval 20 Feet

4. **Remnant Mountains** (Reference: **Wister**, California, Quadrangle)

True desert regions have almost no surface water and no permanent streams, except for those that originate in wetter climates and then flow into the desert. Regardless, water is still important in shaping the landscape and stream courses are easily observed because of the sparse vegetation. An area of southeastern California is depicted on the **Wister** map. This region is in the rainshadow of the Sierra Nevada Mountains and is one of the driest areas in North America. Figure 20.10 is a stereogram representing a section of the **Wister** map showing highly dissected mountains surrounded by eroded debris. There are many temporary stream courses moving weathered rock debris that has covered all but the tops of these mountains.

a. Which direction is the debris moving?

b. Describe the vegetation of this region.

c. Is there evidence of:

    permanent streams? _____

    agriculture? _____

    human occupation? _____

d. Roads are shown on the topographic map. Are they visible in the photographs? _____

e. Note the man-made canal on the lower section of the map. Few people live in this region so what is the likely purpose of the canal?

Figure 20.10   Sierra Nevada Mountains in California. (Photograph courtesy University of Illinois Committee on Aerial Photography and Remote Sensing.)

Wister, California

Scale   1:24,000

Contour Interval 10 Feet

Glaciated Landscapes

Much of the land surface of the mid-high latitudes in both the northern and southern hemispheres have been affected by glacial processes. **Continental glaciers** with thick ice sheets advanced across much of northern Europe, Asia, and North America several times during cold periods of Earth's history. In eastern North America, glaciers extended as far south as the Ohio River valley. High mountain regions throughout the world were occupied by **mountain glaciers** that moved down steep valleys. The scouring of the earth's surface by ice and deposition of glacial sediment created distinct topographic features.

The past 2 million years has been a cold period. This most recent geologic epoch known as the **Pleistocene** ended about 10,000 years ago. During this time there were successive glacial periods interrupted by warm periods when the ice retreated.

Now, glaciers cover only a small percent of the earth's surface compared to most of the Pleistocene. However, most of Antarctica, Greenland, in addition to a few other areas in the high latitudes of the northern hemisphere are still covered by ice. Also, mountain glaciers commonly occur in most of the very high mountain ranges throughout the world. So glacial processes are still shaping the landscape.

1. **Mountain Glacier** (Reference: **Mt. Fairweather**, Alaska, Quadrangle)

    Snow accumulation in high mountain regions will compact into ice, and if thick enough, it will begin to move down the valleys. The ice scours the valley as it moves downslope and therefore, alpine glaciers typically carry large amounts of sediment. Glaciers melt as they move to lower elevations. If the rate of snow accumulation at the source of the glacier high in the mountains exceeds the rate of melting as the glacier descends, the glacier front will advance downslope. Conversely, if the rate of melting exceeds snow accumulation, the glacial front will retreat up the valley.

    Crillion Glacier in Alaska is depicted on the **Mt. Fairweather** map. **Note:** The scale for this map is 1:250,000. The scale for the other topographic maps is 1:24,000. Figure 20.11 is a stereogram showing the lower section of this glacier.

    a. What is the elevation of Mt. Crillion? _____

    b. What causes the dark bands on the glacier? (Hint: The dark band that extends from the point where the north and south glaciers merge.)

    c. A delta has developed at the end of the glacier. Explain the development of this delta.

Figure 20.11 Crillion Glacier in Alaska. (Photograph courtesy University of Illinois Committee on Aerial Photography and Remote Sensing.)

Mt. Fairweather, Alaska

Contour Interval 250 Feet

d. The end of a smaller glacier in a higher valley can be seen on the photograph (C.3-2.5). Is there evidence of melt-water flowing from this glacier? Explain.

e. Ice has plastic characteristics and the glacier can change shape as it flows downslope. As it moves, the surface of the glacier will fracture causing crevasses to develop. Give the photograph coordinates for crevasses present on the Crillion Glacier.

———————————

2. **Glacial U-Shaped Valley** (Reference: **Mt. Moran**, Wyoming, Quadrangle)

Erosion by mountain glaciers deepen and widen the valley floor. Streams will erode valleys with a cross section that is V-shaped. In contrast, mountain glaciers create U-shaped valleys. After the glacier retreats, streams will reoccupy the valley which is sometimes referred to as a **glacial trough**.

A U-shaped valley once occupied by a glacier in the Grand Teton National Park is depicted on the **Mt. Moran** map. Figure 20.12 is a stereogram showing the valley.

Note: Getting a clear stereo image of this valley is difficult. Try looking first at the lower part of the photographs. Be patient and you will get a very good view of the valley.

a. Give the elevation for:

Cascade Creek: _____

Lake of the Crags: _____

Mt. St. John: _____

b. Is there snow in the photograph? Where?

c. Is snow melt-water flowing into the valley? Describe these streams.

d. Where does vegetation occur?

Figure 20.12 Grand Teton National Park in Wyoming. (Photograph courtesy University of Illinois Committee on Aerial Photography and Remote Sensing.)

306

Mt. Moran, Wyoming

Scale   1:24,000

Contour Interval 80 Feet

e.   How has debris flow (landslides) affected the in the bottom of the valley?

3.   **Surface Erosion by a Continental Glacier** (Reference: **Daley Bay**, Minnesota, Quadrangle)

During the Pleistocene, thick continental ice sheets moved southward across Canada and the bordering area of the U.S. near the Great Lakes, including Minnesota, Wisconsin, Michigan, and New York. The glaciers scoured the landscape removing most soil and quarrying out surface rock. This process disrupted drainage patterns and left depressions. As a result, there are numerous lakes over a large area of central North America.

The **Daley Bay** map depicts an area in Minnesota with many lakes occupying depressions caused by glaciation. Figure 20.13 is a stereogram of a small portion of this map.

a.   Grooves in rock made by the last glacier can be seen on the photograph near A.8-3.6. This indicates a movement of ice from either the left or right. Based on these grooves, which direction did the last glacier move? (hint: Do not assume the top of the photograph is north. Refer to the topographic map to determine direction.)

_____

b.   What is the total relief of the area depicted in the topographic map?

_____

c.   Is there evidence of human occupation? Explain.

d.   Do you expect this area to support intensive agriculture? Explain.

Figure 20.13　Lake Kabetogama in Minnesota. (Photograph courtesy University of Illinois Committee on Aerial Photography and Remote Sensing.)

Daley Bay, Minnesota

Scale   1:24,000

Contour Interval 10 Feet

4. **Drumlins** (Reference: **St. Peter**, Wisconsin, Quadrangle)

Drumlins are small, elongated hills composed of unsorted glacial till. The long axis of the drumlins are parallel to the direction of glacial movement. Their form offers minimum resistance to the movement of glacial ice. But the exact formation processes are unknown.

The **St. Peters Bay** map depicts an area in Wisconsin with many drumlins. Figure 20.14 is a stereogram of a small portion of this map.

a. Based on the axis of the drumlins, in what direction was the glacier moving?

        _____

b. What is the elevation of the summit of these drumlins?

average: _____

minimum: _____

maximum: _____

c. Is there evidence that at the time these photographs were taken that these drumlins are severely eroding? Explain.

d. How has this landscape been modified by human activity?

Figure 20.14   Drumlins in Wisconsin. (Photograph courtesy University of Illinois Committee on Aerial Photography and Remote Sensing.)

St. Peters Bay, Wisconsin

Scale   1:24,000

Contour Interval 10 Feet

Aerial Photograph Interpretation and Landform Analysis: Internet Resources

Repeat Photography

http://edcwww.cr.usgs.gov/earthshots/slow/tableofcontentsmap This site from the U.S. Geological Survey has aerial photographs taken at different times of the same location. This is commonly referred to as repeat photography. Repeat photography is often used to show how the environment is changing. Note that on this site you can zoom in/out of photographs and there is an accompanying text with ground level photographs.

Your lab instructor will select **lakes** in different regions of the world for you to view. Note to what extent and why these lakes have changed during the past few decades.

1. Lake: _____ Lat. _____; Long. _____

Dates of Photographs:
1st: _____
2nd: _____
3rd: _____

Changes:

2. Lake: _____ Lat. _____; Long. _____

Dates of Photographs:
1st: _____
2nd: _____
3rd: _____

Changes:

3. Lake: _____   Lat. _____ ; Long. _____

Dates of Photographs:
1st: _____
2nd: _____
3rd: _____

Changes:

Landscape Satellite Images

http://earth.jsc.nasa.gov/ From this NASA site you can search **Earth Landscapes** for images of different types of landforms.

http://earthrise.sdsc.edu/earthrise/main.html This site allows you to zoom in on different regions of the earth and then select photographs taken from the Space Shuttle.

Locate and Order Aerial Photographs

http://www.usgs.gov/ The U.S. Geological Survey provides a site to search and order aerial photographs that cover the entire lower 48 states. Select **Photofinder** to find and order maps for the USGS National Aerial Photography Program.

Appendix A

Trigonometry Tables

Deg.	Sin	Cos	Tan	Cot	Deg
01	.0175	.9998	.0175	57.290	89
02	.0349	.9994	.0349	28.636	88
03	.0523	.9986	.0524	19.081	87
04	.0698	.9976	.0699	14.300	86
05	.0872	.9962	.0875	11.430	85
06	.1045	.9945	.1051	9.514	84
07	.1219	.9925	.1228	8.144	83
08	.1392	.9903	.1405	7.115	82
09	.1564	.9877	.1588	6.313	81
10	.1736	.9848	.1736	5.671	80
11	.1908	.9816	.1944	5.144	79
12	.2079	.9781	.2126	4.704	78
13	.2250	.9744	.2309	4.331	77
14	.2419	.9703	.2493	4.010	76
15	.2588	.9659	.2679	3.732	75
16	.2756	.9613	.2867	3.487	74
17	.2924	.9563	.3057	3.270	73
18	.3090	.9511	.3249	3.077	72
19	.3256	.9455	.3443	2.904	71
20	.3420	.9397	.3640	2.747	70
21	.3584	.9336	.3839	2.605	69
22	.3746	.9272	.4040	2.475	68
23	.3907	.9205	.4245	2.355	67
24	.4067	.9135	.4452	2.246	66
25	.4226	.9063	.4663	2.144	65
26	.4384	.8988	.4877	2.050	64
27	.4540	.8910	.5095	1.962	63
28	.4695	.8829	.5317	1.880	62
29	.4848	.8746	.5543	1.804	61

Deg.	Sin	Cos	Tan	Cot	Deg
30	.5000	.8660	.5774	1.732	60
31	.5150	.8572	.6009	1.664	59
32	.5299	.8480	.6249	1.600	58
33	.5446	.8387	.6494	1.539	57
34	.5592	.8290	.6745	1.482	56
35	.5736	.8192	.7002	1.428	55
36	.5878	.8090	.7265	1.376	54
37	.6018	.7986	.7536	1.327	53
38	.6157	.7880	.7813	1.279	52
39	.6293	.7771	.8098	1.234	51
40	.6428	.7660	.8391	1.191	50
41	.6561	.7547	.8693	1.150	49
42	.6691	.7431	.9004	1.110	48
43	.6820	.7314	.9325	1.072	47
44	.6947	.7193	.9657	1.035	46
45	.7071	.7071	1.000	1.000	45

Appendix B

Diameter and Corresponding Circumference and Area

Diameter (cm)	Circumference (cm)	Area (m²)
3	9.42	.001
4	12.57	.001
5	15.71	.002
6	18.85	.003
7	21.99	.004
8	25.13	.005
9	28.27	.006
10	31.42	.008
11	34.56	.010
12	37.70	.011
13	40.84	.013
14	43.98	.015
15	47.12	.018
16	50.26	.020
17	53.41	.023
18	56.55	.025
19	56.69	.028
20	62.83	.031
21	65.97	.035
22	69.12	.038
23	72.26	.042
24	75.40	.045
25	78.54	.049
26	81.68	.053
27	84.82	.057
28	87.96	.062
29	91.11	.066
30	94.24	.070
31	97.38	.075

32	100.53	.080
33	103.67	.088
34	106.81	.090
35	109.95	.096
36	113.09	.101
37	116.23	.107
38	119.38	.113
39	122.52	.119
40	125.66	.125

Appendix C

Internet Sites

Lab 2   Latitude, Longitude, and Time

Time

http://www.worldtimezone.com/
http://www.time.gov/
http://timeanddate.com

Map Projections

http://hum.aum.edu.pl/~zbzw/glob.glob1.html

Calendars

http://serendipity.magnet.ch/hermetic/index.html
http://www.ecben.net/calendar.shtml

Lab 3   Earth-Sun Relations

Time for Sunrise/Sunset and Solstices and Equinoxes

http://www.usno.navy.mil/
http://www.sunrisesunset.com
http://nws.noaa.gov/er/box/equinox.html

Daylight and Darkness on Earth

http://www.fourmilab.ch/earthview/vplanet.html
http://www.worldtimezone.com/

Lab 4   Solar Radiation

Temperature Anomalies

http://wwwghcc.msfc.nasa.gov/temperature/

Global Warming

http://www.epa.gov/globalwarming/index.html

Lab 5   Atmospheric Pressure and Winds

Surface Pressure and Winds

http://weather.colorado.edu/pages/mainnat.html
http://www.ems.psu.edu/wx//index.html
http://www.nws.noaa.gov/

Lab 6   Atmospheric Humidity

Humidity

http://www.ems.psu.edu/wx/usstats

Weather Satellite Images

http://www.people.fas.harvard.edu/~dbaron/sat/
http://www.cira.colostate.edu/

Lab 8   Air Masses, Cyclonic Storms, and Fronts

U.S. Weather Maps

http://nws.noaa.gov/
http://weather.colorado.edu/pages/mainnat.html

Atlantic Hurricanc Tracking, Maps, and Images

http://fermi.jhuapl.edu/hurr/index.html
http://www.ncdc.noaa.gov/ol/climate/severeweather/hurricanes.html
http://www.nhc.noaa.gov/
http://www.weather.unisys.com/hurricane/atlantic/index.html
http://photolib.noaa.gov/

Lab 9   Weather Analysis

Climate Prediction

http://www.cpc.ncep.noaa.gov/index.htm

Droughts

http://enso.unl.edu/monitor/monitor.html

Lab 10   Climate Classification

Climate Data

http://www.worldclimate.com

Lab 11   Topographic Map Interpretation

Topographic Maps

http://www.topozone.com
http://www.usgs.gov/

Lab 13   Contour Mapping

Elevation Maps of the United States

http://fermi.jhuapl.edu/states/states.html

Lab 14   Soil Properties

Soil Properties

http://www.nrcs.usda.gov/
http://www.statlab.iastate.edu/cgi-bin/osd/osdname.cgi

Lab 15   Soil Erosion

Soil Erosion

http://www.nrcs.usda.gov/

Lab 17   Runoff and Infiltration

Changing Landuse

http://www.nrcs.usda.gov/
http://edcwww.cr.usgs.gov/earthshots/slow/tableofcontentsmap

Water Quality

http://www.epa.gov/surf2/locate/

Lab 18   Stream Discharge

River Stage and Discharge

http://www.usgs.gov/

River Forecasting

http://www.nws.noaa.gov/
http://www.riverwatch.noaa.gov/main.html

Flood Satellite Images

http://edcwww2.cr.usgs.gov/moberly.html
http://www.geo.mtu.edu/rs/tm/flood/

Lab 20   Aerial Photograph Interpretation and Landform Analysis

Repeat Photography

http://edcwww.cr.usgs.gov/earthshots/slow/tableofcontentsmap

Landscape Satellite Images

http://earth.jsc.nasa.gov/
http://earthrise.sdsc.edu/earthrise/main.html

Locate and Order Aerial Photographs

http://www.usgs.gov/

Index

abandoned meander, 273
acute triangles. *see* Solution for the Area of
    Oblique Triangles
adiabatic processes
    adiabatic cooling and heating, defined, 65
    convection, 67–72
    orographic precipitation, 65–67
aerial photograph interpretation, 261–263
    Internet resources about, 317–318
    stereoscopic vision, 263, 264
air masses
    Internet resources about, 87–89
    weather station symbols showing, 73–74
air, rising of. *see* adiabatic processes
albedo (A), 35
alluvial fans, 285–288
altitude
    atmospheric pressure and, 43–45
    of the sun, 27–30
analemma, 22, 31
angular velocity, 21
Antarctic Circle, 26
anticyclones, 47
Arctic Circle, 26
area, 127–128, 141–146, 321–322. *see also* basal
    area; surface area
arid landscapes, 285
    alluvial fans, 285–288
    dissection of a plateau, 293–296
    meandering stream and headward erosion,
        289–292
    remnant mountains, 297–300
Atlantic hurricane tracking, 88–89
atmospheric humidity, 55–61, 63
atmospheric pressure
    altitude and, 43–45
    Coriolis effect, 47
    Internet resources about, 53
    isobars, 46, 48
    winds and, 46–53

average discharge, of streams, 238
azimuth, 1–2

Ballot's law, 48
bankfull, 233–234
barometric pressure (mb), 44, 48–50
basal area
    defined, 198–199
    worksheets for, 213–216
braided streams, 281–284
bulk density (BD), 171

calendars, 20, 323
capacity, 55
central meridians, 13
channel incision, 277
circle of illumination, 26
circumference, 321–322
clay, 167–170, 221, 223–225
climate classification, 99–100
    Cool Temperate climates (D), 99–101, 115–120
    Dry climates (B), 99–101, 103–104, 111–114
    Internet resources about, 119–120
    Koppen Climate Classification System, 101
    Polar climates (E), 99–101
    procedure for, 102–106
    Tropical Humid climates (A), 99–101, 104,
        107–110
    Warm Temperate climates (C), 99–101,
        115–120
climate prediction, 97
cloud formation, 65
compass, 1–2
condensation, 65
continental glaciers, 301, 309–312
contour mapping
    contour interval, 128
    contour line characteristics, 128–129
    data collection for, 151–152
    Internet resources for, 163

plotting for, 153–154
sample data for, 155–156
convection, 67–72
conversion of units, 122–124
Cool Temperate climates (D), 99
    description and criteria for, 101
    map of, 100
    subtypes, 115–120
Coriolis effect, 47
cutoff, 273
cyclones, 47
cyclonic storms. *see also* surface weather maps
    in mid-latitude climates, 118
    Internet resources about, 87–89
    weather station symbols showing, 73–74

day, length of, 23–27, 34
declination of the sun (LS), 22, 31
density
    defined, 197–199
    worksheet for, 212
Desert climates (BW), 111
dew point, 57–58
dew point lapse rate (DPLR), 67
diameter, 321–322
dissection of a plateau, 293–296
distance. *see also* aerial photograph
        interpretation
    calculating with right triangles, 142–143
    measuring, 2
    Representative Fraction (R.F.) scale for, 122
dominance
    defined, 198–199
    worksheet for, 212
droughts, 97–98
drumlins, 313–316
dry adiabatic rate (DAR)
    defined, 65
    in convection, 67–72
    in orographic precipitation, 65–67
Dry climates (B), 99
    description and criteria for, 101
    map of, 100
    subtypes, 103–104, 111–114

Earth
    altitude of the noon sun (ANS) and, 27–30
    declination of sun (LS) and, 22–23, 31
    Internet resources about, 33–34
    length of day/night of, 23–27, 34
    revolution and axis position of, 23
    rotation of, 21–22

elevation. *see* contour mapping; Solutions for
        Right Triangles; topographic map
        interpretation
elevation, water surface, 252
eluviation, 165–166
entrenchment, 277
environmental lapse rate (ELR), 65
equator, 9
equilibrium, 269
erosion. *see* soil erosion; stream discharge;
        individual types of landscapes

field mapping
    compass and measuring azimuth, 1–2
    field notes form for, 4
    pacing, 2
    triangulation, 2–3
flood recurrence intervals, 252–257
flood satellite images, 244–245
floodplains, 269
fluvial processes, 265
    braided streams, 281–284
    lateral erosion, 269–272
    meandering streams, 273–277
    stream incision, 265–268
    stream rejuvenation, 277–280
forest stands. *see also* vegetation analysis
    maps, 202
fronts. *see also* surface weather maps
    in mid-latitude climates, 118
    Internet resources about, 87–89
    weather station symbols showing, 73–74

geographic grid systems, 9. *see also* maps
    coordinates, 10
    used in topographic maps, 121
glacial troughs, 305
glacial U-shaped valleys, 305–309
glaciated landscapes, 301–316
    glacial U-shaped valleys, 305–309
    mountain glaciers, 301–305
    surface erosion by a continental glacier, 309–
        312
global warming, 42
gully development, 192–194

headward erosion, 289
heat. *see* solar radiation
height. *see* Solutions for Right Triangles
horizons, soil, 165–166
humidity. *see* atmospheric humidity
Hydraulic Radius (R), 235–236

illuviation, 165–166
infiltration
    aerial photograph of, 221, 222
    Internet resources about, 227–229
    soil texture and, 221, 223–225
intensity of radiation, 37–40
international date line, 17–18
Internet resources, 87–89, 323–326
    aerial photograph interpretation, 317–318
    air masses, cyclonic storms, fronts, 87–89
    atmospheric humidity, 63–64
    climate classification, 119–120
    climate prediction, 97
    contour mapping, 163
    droughts, 97–98
    landform analysis, 317–318
    latitude, longitude, time, 19–20
    runoff and infiltration, 227–229
    soil erosion, 195
    solar radiation, 41–42
    stream discharge, 243–245
    topographic maps, 139
Intertropical Convergence Zone (ITCZ), 107
isobars, 46, 48

Koppen, W., 99–101

land-use patterns, 227
landform analysis. see also aerial photograph
        interpretation
    arid landscapes, 265, 285–300
    fluvial processes, 265–284
    glaciated landscapes, 265, 301–316
    Internet resources about, 317–318
lateral erosion, 269–272
latitude, 9–12
    Internet resources for, 19–20
    latitudinal position (LP), 28
linear velocity, 21
loams, 167, 223–225
longitude
    defined, 9–12
    Internet resources for, 19–20

magnetic north, 1
Manning's formula, 234
map scale, 122
maps. see also geographic grid systems; Internet
        resources; topographic map interpretation
    barometric pressure, 47
    contour mapping, 128–129, 151–156, 163
    field mapping, 1–4

    for vegetation analysis, 201, 202, 207, 208, 219,
        220
    Mercator Projection, 10, 12, 16
    of Koppen's Climate Classification System, 100
    of United States, 49, 51
    scale of, 122
    surface weather, 75–85
March equinox, 22
meander, 269
meander scars, 273
meandering streams, 273–277, 289–292
Mercator Projection map, 10, 12, 16
meridians, 9. see also longitude
Mid-Latitude climates (C and D), 115–120. see
        also Cool Temperate climates (D); Warm
        Temperate climates (C)
millibar (mb), 43
mountain glaciers, 301–305

National Hurricane Center, 89
National Oceanic and Atmospheric
        Administration (NOAA), 89, 97
National Weather Service, 53
    Internet site of, 87
    River Forecasting Centers, 244
    stations, 49
    weather station symbols, 73–74
night, length of, 23–27, 34
noon sun, altitude of the (ANS), 27–30
north pole, 1

oblique/obtuse triangles. see Solution for the Area
        of Oblique Triangles
orographic precipitation, 65–67
oxbow lakes, 273

pacing, 2
parallels, 9. see also latitude
photographs. see aerial photograph interpretation
plane of the ecliptic, 22
plant cover factor (C), 188–189
plateau, dissection of, 293–296
Pleistocene Epoch, 301, 309
Polar climates (E), 99
    description and criteria for, 101
    map of, 100
pressure, 43. see also atmospheric pressure
pressure gradient, 46
Prime Meridian, 9
psychrometers, 58

radiation. *see* solar radiation
radiation, intensity of, 37–40
rainfall erosion index, 187
recurrence intervals (RI), 247–248, 257–258
    flood frequency curve, 253
    flood recurrence intervals, 252–257
    historical data table, 259
    precipitation curves, 249, 251
    precipitation recurrence intervals, 248–252
    probability of, 248
rejuvenation, 277
relative density, 197–199
relative dominance, 198–199
relative humidity (RH), 55–61, 65
relief, 129
remnant mountains, 297–300
repeat photography, 317–318
Representative Fraction (R.F.), 122, 261, see also
    individual topographic map exercises
right triangles. *see* Solutions for Right Triangles
rise, 134, 236
River Forecasting Centers, 244
rotational velocity, 21–22
Roughness Coefficient (n), 236–237
run, 134, 236
runoff, 221
    aerial photograph of, 221, 222
    infiltration and, 221, 223–225
    Internet resources about, 227–229

sand, 167–170, 221, 223–225
saturated adiabatic rate (SAR), 65
Savannas (Aw), 107
September equinox, 22
silt, 167–170, 221, 223–225
size structure analysis, 203
size-class distribution, 203–204
sling psychrometers, 58
slope factor, 188
slope gradient, 130
slope steepness, 188
soil erodability factor (K), 187, 188
soil erosion, 183–186
    estimating soil loss, 187–189, 191
    gully development and, 192–194
    Internet resources about, 195
    predicting future soil loss, 189–190
    soil erodability factor (K), 187, 188
    universal soil-loss equation for, 183
    weight of soil loss and, 194
soil properties
    bulk density (BD) and, 171

horizons, 165–166
particle size and surface area of, 172
soil profile data sheets, 173, 175
structure of, 167
texture of, 167–170, 221–225
solar angle on the ground (SA_g), 39
solar day, 21
solar radiation, 35–37. *see also* sun
    intensity of, 37–40
    Internet resources for, 41–42
Solution for the Area of Oblique Triangles, 143–146
Solutions for Right Triangles, 141–146
spatial distribution, 204–205
specific humidity, 55–56
stabilization, 269
stage, 252
standard meridians, 13
standard sea-level pressure, 43
Standard Time Zones, 13
state of equilibrium, 269
Steppe climates (BS), 111
stereograms, 263, 264
    of arid landscapes, 286, 290, 294, 298
    of fluvial processes, 266, 270, 274, 278, 282
    of glaciated landscapes, 302, 306, 310, 314
stream discharge
    bankfull discharge estimation and, 233–234
    calculation of, 231–232
    erosion potential from, 232–233
    Internet resources about, 243–245
    prediction of, 238–241
    velocity and, 231, 234–238
Stream Gradient (S), 134, 236
structure, of soil, 167
Subtropical climates. *see* Warm Temperate
    climates (C)
summer solstice, 22, 24
sun
    altitude of the, 27–30
    declination of, 22–23, 31
    earth rotation and, 21–22
    Internet resources about, 33–34
    length of day/night and, 23–27, 34
    solar radiation, 35–42
surface area, 172. *see also* Solution for the Area of
    Oblique Triangles
surface erosion by a continental glacier, 309–312
surface weather maps, 75
    August 24/26, 1992 example, 77–79
    February 11, 1988 example, 74–76
    October 16-19, 1990 example, 80–85

symbols
    for topographic maps, 121–122
    for weather stations, 73–74

temperature. *see* solar radiation
textural classes, of soil, 167–168, 221, 223–225
texture, of soil, 167–170, 223–225
time, 13–16
    international date line, 17–18
    Internet resources for, 19–20, 323
topographic map interpretation, 121–122. *see also*
      aerial photograph interpretation; landform
      analysis; *individual types of landscapes*
    calculating area, 127–128
    contours, 128–130
    conversion of units, 122–124
    Internet resources for, 139
    map examples, 125–126, 131–133, 137–138
    map scale, 122
    slope gradient, 130
    Stream Gradient, 134
    symbols for, 121–122
    topographic contours, 128–129
trees. *see* vegetation analysis
triangulation, 2–3
trigonometry applications
    Solution for Area of Oblique Triangles, 143–
      146
    Solutions for Right Triangles, 141–146
    tables for, 319–320
Tropic of Cancer, 22
Tropic of Capricorn, 22
Tropical Humid climates (A), 99
    description and criteria for, 101
    map of, 100
    subtypes, 104, 107–110
Tropical Monsoons (Am), 107
Tropical West climates (Af), 107

U.S. Geological Survey, 243–245, 317–318
unit conversion, 122–124
United States maps, 49, 51
universal soil-loss equation, 183

vegetation analysis
    basal area worksheets for, 213–216
    climate classification and, 99
    density, 197–199, 205, 212
    dominance, 198–199, 212
    forest stand map examples, 201, 202, 207, 208,
      219, 220
    size structure analysis, 203
    size-class distribution, 203–204, 209
    spatial distribution, 204–205
    spatial patterns of species, 210
    surface age, 211
    tree age/growth rates, 210–211, 217
velocity, 231
volume, of gullies, 192–194

Warm Temperate climates (C), 99
    description and criteria for, 101
    map of, 100
    subtypes, 115–120
water quality, 228–229
water surface elevation, 252
water vapor, 55, 56
weather analysis
    data collection for, 91–95
    Internet resources about, 97–98
weather satellite images, 63–64
weight, of soil loss, 194
wet-bulb depression, 58
winds
    barometric pressure and, 48–50
    Coriolis effect on, 47
    Internet resources about, 53
    isobars showing, 47, 48
winter solstice, 22, 24
World Time Zones, 13, 14

Topographic Map Symbols

...DARIES

...al ....
...r territorial ....
... or equivalent ....
...wnship or equivalent ....
...rated-city or equivalent ....
...eservation, or monument ....
...ark ....

...SURVEY SYSTEMS

...blic Land Survey System:
...ship or range line ....
...ation doubtful ....
...n line ....
...ation doubtful ....
...d section corner; found closing corner ....
...ss corner; meander corner .... WC   MC

...and surveys:
...ship or range line ....
...on line ....
...rant or mining claim; monument ....
...ine ....

...AND RELATED FEATURES

...y highway ....
...ary highway ....
...uty road ....
...oved road ....
...
...ghway ....
...ghway with median strip ....
...nder construction ....
...ass; overpass ....
...
...idge ....

...NGS AND RELATED FEATURES

...g or place of employment: small; large ...
...church ....
...arehouse, etc.: small; large ....
...mission tint ....
...ck ....
...
...strip ....
...ther than water); windmill ....
...ank: small; large ....
...ank: small; large ....
...d reservoir ....
...station ....
...rk object ....
...ound; picnic area ....
...ry: small; large .... Cem

RAILROADS AND RELATED FEATURES

Standard gauge single track; station ....
Standard gauge multiple track ....
Abandoned ....
Under construction ....
Narrow gauge single track ....
Narrow gauge multiple track ....
Railroad in street ....
Juxtaposition ....
Roundhouse and turntable ....

TRANSMISSION LINES AND PIPELINES

Power transmission line: pole; tower ....
Telephone or telegraph line ....
Aboveground oil or gas pipeline ....
Underground oil or gas pipeline ....

CONTOURS

Topographic:
   Intermediate ....
   Index ....
   Supplementary ....
   Depression ....
   Cut; fill ....
Bathymetric:
   Intermediate ....
   Index ....
   Primary ....
   Index Primary ....
   Supplementary ....

MINES AND CAVES

Quarry or open pit mine ....
Gravel, sand, clay, or borrow pit ....
Mine tunnel or cave entrance ....
Prospect; mine shaft ....
Mine dump ....
Tailings ....

SURFACE FEATURES

Levee ....
Sand or mud area, dunes, or shifting sand .....
Intricate surface area ....
Gravel beach or glacial moraine ....
Tailings pond ....

VEGETATION

Woods ....
Scrub ....
Orchard ....
Vineyard ....
Mangrove ....

COASTAL FEATURES

Foreshore flat ....
Rock or coral reef ....
Rock bare or awash ....
Group of rocks bare or awash ....
Exposed wreck ....
Depth curve; sounding ....
Breakwater, pier, jetty, or wharf ....
Seawall ....

BATHYMETRIC FEATURES

Area exposed at mean low tide; sounding datum .
Channel ....
Offshore oil or gas: well; platform ....
Sunken rock ....

RIVERS, LAKES, AND CANALS

Intermittent stream ....
Intermittent river ....
Disappearing stream ....
Perennial stream ....
Perennial river ....
Small falls; small rapids ....
Large falls; large rapids ....

Masonry dam ....

Dam with lock ....

Dam carrying road ....

Intermittent lake or pond ....
Dry lake ....
Narrow wash ....
Wide wash ....
Canal, flume, or aqueduct with lock ....
Elevated aqueduct, flume, or conduit ....
Aqueduct tunnel ....
Water well; spring or seep ....

GLACIERS AND PERMANENT SNOWFIELDS

Contours and limits ....
Form lines ....

SUBMERGED AREAS AND BOGS

Marsh or swamp ....
Submerged marsh or swamp ....
Wooded marsh or swamp ....
Submerged wooded marsh or swamp ....
Rice field ....
Land subject to inundation ....

MILE SCALE 1:62 500

UNITED STATES
DEPARTMENT OF THE INTERIOR
GEOLOGICAL SURVEY

TOPOGRAPHIC
MAP INFORMATION AND SYMBOLS
MARCH 1978

QUADRANGLE MAPS AND SERIES

Quadrangle maps cover four-sided areas bounded by parallels of latitude and meridians of longitude. Quadrangle size is given in minutes or degrees.

Map series are groups of maps that conform to established specifications for size, scale, content, and other elements.

Map scale is the relationship between distance on a map and the corresponding distance on the ground.

Map scale is expressed as a numerical ratio and shown graphically by bar scales marked in feet, miles, and kilometers.

NATIONAL TOPOGRAPHIC MAPS

Series	Scale	1 inch represents	1 centimeter represents	Standard quadrangle size (latitude-longitude)	Quadrangle area (square miles)
7½-minute	1:24,000	2,000 feet	240 meters	7½ × 7½ min.	49 to 70
7½ × 15-minute	1:25,000	about 2,083 feet	250 meters	7½ × 15 min.	98 to 140
Puerto Rico 7½-minute	1:20,000	about 1,667 feet	200 meters	7½ × 7½ min.	71
15-minute	1:62,500	nearly 1 mile	625 meters	15 × 15 min.	197 to 282
Alaska 1:63,360	1:63,360	1 mile	nearly 634 meters	15 × 20 to 36 min.	207 to 281
Intermediate	1:100,000	nearly 1.6 miles	1 kilometer	30 × 60 min.	1568 to 2240
U. S. 1:250,000	1:250,000	nearly 4 miles	2.5 kilometers	1° × 2° or 3°	4,580 to 8,669
U. S. 1:1,000,000	1:1,000,000	nearly 16 miles	10 kilometers	4° × 6°	73,734 to 102,759
Antarctica 1:250,000	1:250,000	nearly 4 miles	2.5 kilometers	1° × 3° to 15°	4,089 to 8,336
Antarctica 1:500,000	1:500,000	nearly 8 miles	5 kilometers	2° × 7½°	28,174 to 30,462

CONTOUR LINES SHOW LAND SHAPES AND ELEVATION

The shape of the land, portrayed by contours, is the distinctive characteristic of topographic maps.

Contours are imaginary lines following the ground surface at a constant elevation above or below sea level.

Contour interval is the elevation difference represented by adjacent contour lines on maps.

Contour intervals depend on ground slope and map scale. Small contour intervals are used for flat areas; larger intervals are used for mountainous terrain.

Supplementary dotted contours, at less than the regular interval, are used in selected flat areas.

Index contours are heavier than others and most have elevation figures.

Relief shading, an overprint giving a three-dimensional impression, is used on selected maps.

Orthophotomaps, which depict terrain and other map features by color-enhanced photographic images, are available for selected areas.

COLORS DISTINGUISH KINDS OF MAP FEATURES

Black is used for manmade or cultural features, such as roads, buildings, names, and boundaries.

Blue is used for water or hydrographic features, such as lakes, rivers, canals, glaciers, and swamps.

Brown is used for relief or hypsographic features—land shapes portrayed by contour lines.

Green is used for woodland cover, with patterns to show scrub, vineyards, or orchards.

Red emphasizes important roads and is used to show public land subdivision lines, land grants, and fence and field lines.

Red tint indicates urban areas, in which only landmark buildings are shown.

Purple is used to show office revision from aerial photographs. The changes are not field checked.

INDEXES SHOW PUBLISHED TOPOGRAPHIC MAPS

Indexes for each State, Puerto Rico and the Virgin Islands of the United States, Guam, American Samoa, and Antarctica show available published maps. Index maps show quadrangle location, name, and survey date. Listed also are special maps and sheets, with prices, map dealers, Federal distribution centers, and map reference libraries, and instructions for ordering maps. Indexes and a booklet describing topographic maps are available free on request.

HOW MAPS CAN BE OBTAINED

Mail orders for maps of areas east of the Mississippi River, including Minnesota, Puerto Rico, the Virgin Islands of the United States, and Antarctica should be addressed to the Branch of Distribution, U. S. Geological Survey, 1200 South Eads Street, Arlington, Virginia 22202. Maps of areas west of the Mississippi River, including Alaska, Hawaii, Louisiana, American Samoa, and Guam should be ordered from the Branch of Distribution, U. S. Geological Survey, Box 25286, Federal Center, Denver, Colorado 80225. A single order combining both eastern and western maps may be placed with either office. Residents of Alaska may order Alaska maps or an index for Alaska from the Distribution Section, U. S. Geological Survey, Federal Building-Box 12, 101 Twelfth Avenue, Fairbanks, Alaska 99701. Order by map name, State, and series. On an order amounting to $300 or more at the list price, a 30-percent discount is allowed. No other discount is applicable. Prepayment is required and must accompany each order. Payment may be made by money order or check payable to the U. S. Geological Survey. Your ZIP code is required.

Sales counters are maintained in the following U. S. Geological Survey offices, where maps of the area may be purchased in person: 1200 South Eads Street, Arlington, Va.; Room 1028, General Services Administration Building, 19th & F Streets NW, Washington, D. C.; 1400 Independence Road, Rolla, Mo.; 345 Middlefield Road, Menlo Park, Calif.; Room 7638, Federal Building, 300 North Los Angeles Street, Los Angeles, Calif.; Room 504, Custom House, 555 Battery Street, San Francisco, Calif.; Building 41, Federal Center, Denver, Colo.; Room 1012, Federal Building, 1961 Stout Street, Denver Colo.; Room 1C45, Federal Building, 1100 Commerce Street, Dallas, Texas; Room 8105, Federal Building, 125 South State Street, Salt Lake City, Utah; Room 1C402, National Center, 12201 Sunrise Valley Drive, Reston, Va.; Room 678, U. S. Court House, West 920 Riverside Avenue, Spokane, Wash.; Room 108, Skyline Building, 508 Second Avenue, Anchorage, Alaska; and Federal Building, 101 Twelfth Avenue, Fairbanks, Alaska.

Commercial dealers sell U. S. Geological Survey maps at their own prices. Names and addresses of dealers are listed in each State index.

INTERIOR—GEOLOGICAL SURVEY, RESTON, VIRGINIA—1978

41500031

MILE SCALE 1:24 000

FOOT SCALE 1:24 000

FOOT SCALE 1:62 500